JOURNAL OF CYBER
SECURITY AND MOBILITY

Volume 2, No. 2 (April 2013)

Special issue on

The Mobility Age: Opportunities, Challenges and Solutions

Guest Editor:

Shweta Jain
York College CUNY, USA

JOURNAL OF CYBER SECURITY AND MOBILITY

Editors-in-Chief
Ashutosh Dutta, AT&T, USA
Ruby Lee, Princeton University, USA
Neeli R. Prasad, CTIF-USA, Aalborg University, Denmark

Associate Editor
Shweta Jain, York College CUNY, USA

Steering Board
H. Vincent Poor, Princeton University, USA
Ramjee Prasad, CTIF, Aalborg University, Denmark
Parag Pruthi, NIKSUN, USA

Advisors
R. Chandramouli, Stevens Institute of Technology, USA
Anand R. Prasad, NEC, Japan
Frank Reichert, Faculty of Engineering & Science University of Agder, Norway
Vimal Solanki, Corporate Strategy & Intel Office, McAfee, Inc, USA

Editorial Board

Sateesh Addepalli, CISCO Systems, USA
Mahbubul Alam, CISCO Systems, USA
Jiang Bian, University of Arkansas for Medical Sciences, USA
Tsunehiko Chiba, Nokia Siemens Networks, Japan
Debabrata Das, IIIT Bangalore, India
Subir Das, Telcordia ATS, USA
Tassos Dimitriou, Athens Institute of Technology, Greece
Pramod Jamkhedkar, Princeton, USA
Eduard Jorswieck, Dresden University of Technology, Germany
LingFei Lai, University of Arkansas at Little Rock, USA
Yingbin Liang, Syracuse University, USA
Fuchun J. Lin, Telcordia, USA

Rafa Marin Lopez, University of Murcia, Spain
Seshadri Mohan, University of Arkansas at Little Rock, USA
Rasmus Hjorth Nielsen, Aalborg University, Denmark
Yoshihiro Ohba, Toshiba, Japan
Rajarshi Sanyal, Belgacom, Belgium
Andreas U. Schmidt, Novalyst, Germany
Remzi Seker, University of Arkansas at Little Rock, USA
K.P. Subbalakshmi, Stevens Institute of Technology, USA
Reza Tadayoni, Aalborg University, Denmark
Wei Wei, Xi'an University of Technology, China
Hidetoshi Yokota, KDDI Labs, USA

Aim
Journal of Cyber Security and Mobility provides an in-depth and holistic view of security and solutions from practical to theoretical aspects. It covers topics that are equally valuable for practitioners as well as those new in the field.

Scope
The journal covers security issues in cyber space and solutions thereof. As cyber space has moved towards the wireless/mobile world, issues in wireless/mobile communications will also be published. The publication will take a holistic view. Some example topics are: security in mobile networks, security and mobility optimization, cyber security, cloud security, Internet of Things (IoT) and machine-to-machine technologies.

JOURNAL OF CYBER SECURITY AND MOBILITY

Volume 2   No. 2   April 2013

Published, sold and distributed by:
River Publishers
P.O. Box 1657
Algade 42
9000 Aalborg
Denmark

Tel.: +45369953197
www.riverpublishers.com

Journal of Cyber Security and Mobility is published four times a year.
Publication programme, 2013: Volume 2 (4 issues)

ISSN 2245-1439 (Print Version)
ISSN 2245-4578 (Online Version)
ISBN 978-87-92982-66-7 (this issue)

Editorial Foreword

Welcome to the special issue of the *Journal of Cyber Security and Mobility*. This issue brings together cutting edge research in mobility and enterprise security to celebrate the revolutionary "Mobility Age". Millions of smart devices (smartphones, tablets, Internet of Things, networked vehicles, sensors and countless other low profile, low powered entities) interact with the virtual world to fetch content, perform computation and report the parameters of the physical world. In the future, this suite of devices and applications is set to grow and the ideas will be limited only by imagination.

The Mobility Age comes with its own problems and challenges. First, there are concerns for self-organization of small devices in emerging networks such as vehicular ad-hoc networks (VANET), Internet of Things (IoT) and mobile ad-hoc networks (MANET). The size and dynamics of these networks makes centralized control, communication and computationally extensive, which is why distributed solutions are explored. Second, power and network contention needs to be managed to enable longer network lifetime and efficient operation. Third, traditional networking principles that are based on network addresses and locations, need to be redesigned to fit the dynamics of mobile networking paradigm. Finally, mobility also challenges the security of fixed enterprise assets by launching distributed attacks from fleeting sources that can be hard to locate. Despite the challenges, the mobility age brings with itself the opportunity to implement new ideas that can often be expensive if traditional fixed infrastructure is used.

Since this issue is "celebrating" the Mobility Age, we will start with the positives. The first paper in this issue is by Amitabha Mishra et al., who present a refreshing idea that has the promise of bringing cloud computing "across the digital divide, to developing nations". The authors show how millions of mobile phones can be federated to run micro computations, which when put together, can match the capacity of expensive and high profile data centers. The second paper by Tan Yan et al. presents a similarly revolutionary concept in which interest based information can be relayed to networked vehicles in a VANET, without a priori knowledge of their network address or identity. Novel networking principles are used to achieve the objective with low control messaging overhead and latency while ensuring high data delivery ratio. Next we dive into important issues in the mobility age, the first being the question of self-organization to improve the network lifetime. The third paper by Janusz Kusyk et al. shows how autonomous robots can use a bio inspired game to achieve just that in a mobile ad-hoc network (MANET). The objective is to achieve a uniformly distributed topology so that nodes can operate at low power levels while maintaining connectivity with neighbors and hence achieve longer network lifetime. The second important issue in the mobility age is of security. With millions of smart devices in the network, security becomes an important concern not only for the devices themselves, but also for fixed assets that are now vulnerable to attacks by millions of compromised and malicious smart phones and other such networked systems. The fourth paper by Wang et al. addresses the problem of detecting sophisticated targeted attacks on enterprise systems. The authors present a technique in which a smart attacker is deceived into honey assets, and eventually detected through multiple layers of deception.

We would like to thank the reviewers, editorial board members, advisory board members, steering committee members and the staff of River Publishers for their efforts in preparing the publication of this special issue of the journal. We hope our readers will continue to find the forthcoming issues of this journal useful in enriching their knowledge of cyber security and mobility.

Shweta Jain
York College CUNY, USA

MoCCA: A Mobile Cellular Cloud Architecture

Amitabh Mishra and Gerald Masson

Johns Hopkins University Information Security Institute, Baltimore, MD, USA;
e-mail: amitabh@cs.jhu.edu, masson@jhu.edu

Received 31 May 2013; Accepted 15 July 2013

Abstract

This paper presents MoCCA – a cellular cloud architecture for building mobile clouds using small-footprint micro-servers running on cell phones. We provide details of this architecture which is based on GSM standard, discuss several challenges, and include performance results to validate the assumptions that a mobile cellular cloud can indeed be in the realm of possibilities.

Keywords: Mobile cellular networks, cloud computing, performance.

1 Introduction

Cloud computing is a new computing paradigm, involving data and/or computation outsourcing, with infinite and elastic resource scalability. The NIST defines cloud computing as: "a model for enabling convenient, on-demand network access to a shared pool of configurable computing resources (e.g., networks, servers, storage, applications, and services) that can be rapidly provisioned and released with minimal management effort or service provider interaction" [15].

Cloud computing service providers typically possess large data centers consisting of a large number of servers. The resources of these servers are provisioned to the clients on demand. The cloud service providers typically

Journal of Cyber Security and Mobility, Vol. 2, 105–125.

provide one or more of the following services: Software-as-a-service (SaaS), Platform-as-a-service (PaaS), or Infrastructure-as-a-service (IaaS). In SaaS, the cloud provider runs applications on the cloud platform, and clients usually access these applications via a web-browser interface. Examples of SaaS are Google Docs, Microsoft Live, etc. In PaaS, clients get more access to the cloud, by deploying or configuring applications or code that runs on the cloud providers software and operating system platform. In IaaS, clients get the greatest control over the cloud. In this model, the clients can deploy operating system on virtual machines running on the cloud servers, provision resources, and run arbitrary software. Data Flow programming has gained popularity for data processing in Clouds. A widely used programming framework in this model is Google's MapReduce [5]. Given the operational advantage of using a cloud for computing, they have become widely popular in the developed world. However, if we want to bring cloud computing across the digital divide, to developing nations, we face a big obstacle. As of now, setting up a moderate sized cloud requires significant investment in the data center infrastructure. A high-density data center with 10000 servers can cost up to $4 million or more [20]. Larger scale facilities such as Microsoft Azure's Chicago data center cost $500 million [12, 16]. A large part of this cost goes towards setting up the physical infrastructure for cooling, racks etc., making it prohibitively costly for a new player to enter the cloud computing business.

To solve this problem, we propose using the widely deployed mobile phones (MS) as the building blocks for clouds. We define a mobile cloud as a cloud composed of mobile phones and the associated networking components. In contrast with traditional clouds, a mobile cloud is formed using loosely connected servers. A mobile cloud provides a lot of business advantages over traditional clouds. It can allow base station owners or cell phone operators to enter the market with little establishment cost. Base stations which are integral part of the mobile cloud architecture are already deployed in most places. To build a mobile cloud, a provider only needs to deploy additional software and minimal amount of additional hardware to the base stations. Running a mobile cloud microserver on a cell phone uses up energy and phone resources, so the clients must have some incentives to participate in such a cloud.

Unlike a traditional data center, the mobile cloud does not require extensive cooling systems, a large building for servers, wiring, and that it being close to power utilities. A mobile cloud is also self-balancing when used to serve local computing needs. For example, in a building with few cell phone users, the resources required to run a local service is low. So, a base station can run local services on the few cell phones present in the building. If more cell phones enter the base station's service area, the resource requirements increase accordingly, but at the same time, the base station has access to more cell phones to scale its services.

In this paper, we propose a Mobile Cellular Cloud Architecture (MoCCA) – that aims at building cloud computing systems using smart mobile phones. Unlike traditional clouds that depend on co-located servers connected via a local area network, we propose building a cloud using loosely-connected micro-servers – small footprint code that runs on a cell phone – connected via a wireless link to a base station. The contributions of this paper are as follows: (1) Identification of challenges in building mobile clouds; (2) Introduction of a churn tolerant cloud computing architecture that uses microservers running on mobile phones as compute nodes and facilitates building an autonomic cloud over wireless cellular networks. (3) Demonstration using an analytic performance model that a cellular cloud is indeed feasible. The rest of this paper is organized as follows: Section 2 presents issues and challenges in building a cloud with mobile phones. The details of MoCCA are presented in Section 3. Section 4 includes the performance of a set of numerical applications suitable for cloud computing. We discuss the state of the art/related work in mobile cloud computing in Section 5 followed by the conclusions.

2 Challenges

A mobile cloud introduces a set of new challenges in terms of operation and security of the cloud components, as well as the data objects. The challenges mainly arise from the fundamental differences between the network topology and node characteristics of a wired data center network and a cellular mobile network. In this section, we discuss the operational and security related challenges in designing a mobile cloud.

2.1 Operational Issues

Connectivity: The biggest issue facing a mobile cloud is connectivity. Data centers are built using high-speed wired networks, operating under the control of the cloud service provider. In contrast, mobile clouds, built using small-footprint server code running on mobile phones, do not have fast and high-bandwidth connectivity. Unlike a data center connected via a local area network, mobile phones use the star topology where a large number of phones are connected to a base station. The phones do not communicate among themselves directly rather they connect through the base station. This difference in topology implies that we need solutions that allow cloud protocols to deal with frequent network disruptions.

Computational limitation: Mobile phones do not possess powerful computational capability. Unlike traditional cloud servers with multi-core, high-speed processors, mobile phones may not be as resource rich. Hence, mobile clouds must be designed with the computational limitations in mind. The task assigned to each of the mobile cloud nodes should be such that it is within its resource constraints such as CPU, memory, bandwidth and the energy.

Churn: A mobile cloud needs to consider churn in a large scale. Since mobile phone users are not limited to a single location and may move in and out of range of a given cellular base station, there is a high chance that a mobile cloud node will drop out, run out of power, or have the server process interrupted or terminated. The mobile cloud architecture and protocols must be designed with enough redundancy to handle such large scale churn.

Energy: Mobile phones are power limited devices. Computations and data transfer will drain the battery. Hence, the amount of data assigned to each mobile node needs to be small enough to be feasible to compute and transfer efficiently.

User incentive: Since the nodes in the mobile cloud are mobile phones owned by different people, we need to provide an incentive model for the users to allow the use of their phones. Specifically, the users need to have a compensation model where they will benefit from letting their unused phones be a part of a mobile cloud. One possibility can be that the cell phone company may provide a discount or monetary incentive in return of the use of phone CPU cycles and data bandwidth. Since, cell phones are typically unused most of the day, especially during off-peak hours. So, the cell phone owners can benefit financially by letting the phone company utilize the unused cell CPU cycles.

2.2 Security and Privacy Issues

We mention these issues here rather briefly to appraise readers to security challenges that mobile cloud computing faces. But, security issues are beyond the scope of this paper.

Confidentiality and privacy: Unlike traditional cloud servers, the individual mobile phones in a mobile cloud are owned by different people, some of whom may not be trustworthy. Hence, it is vital to ensure that attackers cannot gather confidential information by pretending to be a legitimate user. Even if the adversary takes over a phone, or infiltrates the cloud, she should learn only small pieces of data (that was sent to her phone) and nothing else about the whole computation. Also, the cloud server code running on the phone should not gather sensitive personal information from the user's phone. Another challenge is to ensure the security of data in transit – in mobile clouds, most of the data transfer will occur over wireless links, which brings in the possibility of eavesdropping attacks.

Integrity: Mobile clouds need to ensure the integrity of data as well as computations. Dishonest users can try to benefit by misreporting the result of a computation (perhaps by sending random results without performing the actual computation). We must also ensure that only the intended data processing functions (either standard or user-defined) were applied to produce the results.

Forensics: A mobile cloud must have strong support for digital forensics. As the nodes are no longer owned by the service provider, the risk of attacks is higher. Hence, we need schemes for logging, provenance management, and forensics to identify misbehavior.

3 MOCCA: Mobile Cellular Cloud Architecture

To build clouds using mobile phones, we introduce the *Mobile Cellular Cloud Architecture (MoCCA)*. On a high-level, MoCCA is based on a small-footprint microserver code running on mobile phones. These microservers are co-ordinated by a management process running in the base station. In this section, we present an overview of MoCCA which assumes GSM cellular network as the basis for the mobile cloud. We begin by discussing background information on GSM network topology.

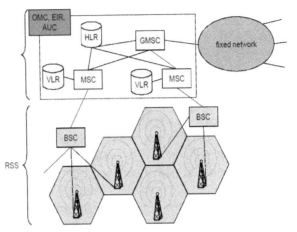

Figure 1   Architecure of a GSM cellular system [17].

3.1 Background

GSM is a widely used narrowband digital cellular system belonging to the 2nd generation (G) and has further evolved to GPRS (General Packet Radio Service – 2.5G), and UMTS (Universal Mobile Telecommunications System – 3G) systems. A functional architecture of GSM [17] is shown in Figure 1. Here, each base station is shown as a hexagon which serves N number of mobile stations (MS). A base station controller (BSC) controls multiple base stations and interfaces to a mobiles switching center (MSC). For location and mobility management of mobile devices Home Location (HLR) and Visitor Location (VLR) Registers are used [17].

3.2 MoCCA Overview

In MoCCA, the mobile cloud comprises of smart phones, base stations (BS), base station controllers (BSC), and mobile switching centers (MSC). A small mobile cloud can be formed out of mobile stations served by a single base station. A medium cloud can be constructed out of phones served by multiple base stations within the control of the same BSC. A larger cloud can be formed out of coverage area supported by multiple BSCs that are under the control of one MSC. A much larger cloud can still be constructed at the PSTN (Public Switch Telephone Network) level that includes multiple MSCs. Because of the hierarchical nature of cellular networks, MoCCA is inherently scalable.

We make the following assumptions in order to develop MoCCA: (1) The BSC has the ability to partition the client initiated workload into multiple smaller chunks that can be processed by a multiple of smart phones. (2) The BSC has a list of smart phones that are willing to participate in sharing the workload assigned to the cloud. (3) Using the control plane (existing signaling architecture), a BSC can allow a phone to join or leave the cloud as and when necessary. However, for better performance, it is reasonable to expect that a phone participates in a cloud for a minimum amount of time that is agreed upon during the initialization. (4) Phones have required software to compute the results, or it could be downloaded from the base station or any other appropriate facility. (6) After completion of the assigned job, the mobile phone informs the base station that the job is completed using a message on a control channel and transmits the result on one of the data channels assigned by the base station. (7) The BSC is responsible for verifying the correctness of the results received from the multiple phones.

3.3 Operational Model

MoCCA uses a dataflow model of cloud computing, such as MapReduce [5]. In such data processing systems, each node performs a mapping or reducing function. Results from one stage are fed into the next stage. We consider data-parallel computations.

To submit a job to MoCCA, a cloud client contacts a BSC and sends it the data and the specification of the code that should operate on the data (e.g., map and reduce functions). The BSC then contacts the MoCCA microservers running on the mobile phones, through the MoCCA manager software running on the base stations.

MoCCA microserver: Each mobile phone runs a MoCCA microserver process that handles communication with the base station and management of the server function and data. A controller component in the microsever manages the communication with the MoCCA manager, retrieval and deployment of function code and data, maintenance of accounting information, and the return of the results to the BS. The code is executed in a sandbox (similar to Java Applets), and is prevented from accessing any resource outside the sandbox. This ensures that any malicious code will not be able to read any . personal information stored on the phone.

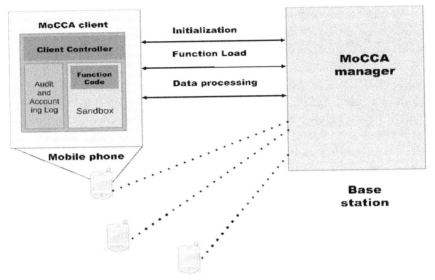

Figure 2 Operational model of MoCCA.

Operation: On a high level, the communication between a mobile phone and the base station happens in three phases.

1. *Initialization phase*: When a mobile wants to join the MoCCA framework, it sends a join request to the base station controller. The BSC then initiates an authentication protocol that is based on challenge and response methodology. To further strengthen this protocol, after the phone is authenticated, the BS and the MS can perform a Diffie–Hellman key exchange to establish a cryptographic session key. This key is used to encrypt any future communication between the cell phone and the base station. In this phase, the MS may also negotiate the type of service it is willing to provide to the BSC including the amounts of data it will process and the number of sessions for which it will remain active. Accounting for the service is also initiated in this phase.

2. *Function load phase*: In this phase, the MoCCA manager in the BS sends function code to the phone. The code is placed in the microserver sandbox.

3. *Data processing phase*: In this phase, the BSC sends one or more sets of data to the phone via BS. On receiving a data object, the MoCCA microserver places the data object in the sandbox, and invokes the previously loaded function to operate on the data object.

After the end of each computation, the microserver controller sends the result and accounting information back to the base station. The operational model of MoCCA is shown in Figure 2.

3.4 Data and Control Plane Operations

In MoCCA, all cloud computing messages whether related to control or data plane are handled on designated control and data channels thereby suggesting minimum or no modifications to existing GSM system.

3.5 Performance and Reliability

Bandwidth: The bandwidth provided by emerging cellular systems for data applications has been increasing over the years due to the emergence of bandwidth hungry mobile applications such as mobile web-surfing and multimedia. At the present time GSM/Edge systems provide a bandwidth of 384 Kbps which increases to 2 Mbps for UMTS/DECT systems. Emerging 4 G systems such as LTE are planned to have downlink bandwidth of 84–168 Mbps [24].

Energy: For cloud applications running on mobile phones, energy consumption in computation and communication (transmission and reception) is of paramount importance. For example, in GSM based applications running on Android G1 smart phones, the cost of transmission of a 200 byte data packet is 4.67 Joules while the cost of reception is 2.05 Joules [2]. The energy cost of computation is several orders of magnitude less than wireless communications. The average energy capacity of Android G1 battery is approximately 15000–20000 Joules. These results justify our hypothesis that mobile phones can indeed be appropriate platforms to act as servers for mobile clouds.

it Frequent Connection/Disconnection: In cellular communication, the probability of connection impairments leading to eventual disconnection is very high when a cloud server node is mobile. Also there is another important difference that is related to mobile cloud. In a cellular voice or data applications at the termination of the session, no data transmission takes place. But in the case of mobile cloud, the phone is responsible for sending back the computed results to the original BSC completing the task. There are several issues that arise in this context such as:

1. A MS acting as a cloud server is mobile within the coverage area of the serving base station and does not cross the cell boundary; so no handoffs take place. If there is a disconnection while computed results are being transferred to the base station, or software is being downloaded from the base station, or input data is being transmitted by the base station, then these items (text and/or data) need to be retransmitted by the base station or phone using the existing mechanisms in place for cellular data communications or as specified in standards.

2. The second scenario arises when a phone acting as a cloud server moves to a different cell resulting in a handoff or series of handoffs. We assume computed results will reach the originating base station following the handoff trail.

3.6 Reassembly and Correctness

The base station controller will be assigned the task of reassembling the results from different smart phones and ensuring that the transmission errors on the wireless channels have not introduced any errors in the computed results. The assumptions that we make in MoCCA is that cellular network architecture and governing specifications do not change to accommodate mobile cloud computing. Cellular networks implement forward error correction to deal with the channel errors and this will be first line of defence with regard to transmission errors. In addition in MoCCA, we introduce another feature to ensure correctness of the received results based on the theory of Triple Modular Redundancy (TMR) proposed by Von Neumann [18]. In TMR identical computations are assigned to three smart phones outputs of which are compared at the base station controller through a voter which chooses the correct result. A TMR configuration is shown in Figure 3. Here the duplication with output comparison is considered as an error detection technique.

The availability of the third copy of the computation provides enough redundant information to allow error masking in any one of the three copies. This is accomplished by means of a majority (two-out-of-three) vote on the three copies of the computed results. The reliability of the results of this configuration can be given as

$$R = R_v \times (R_m^2 + 3R_m^2(1 - R_m)), \tag{1}$$

where R_v and R_m are the reliabilities of the voter and a single copy of triplicated computations. The concept of triple modular redundancy can be extended to include N copies with majority voting at the base station con-

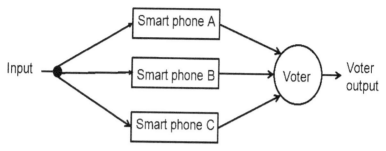

Figure 3 Reassembly and correctness in MoCCA.

troller. Equation (1) can be extended to N modular redundancy scheme to Equation (2), if higher degree of reliability in computed results is required.

$$R = R_v \sum_{i=0}^{N/2} \binom{N}{i} R_m^{(N-1)} + (1 - R_m)'. \tag{2}$$

As part of the research agenda, we will examine what is the optimal redundancy that is required to guarantee the correctness of the results and its energy and transmission cost implications. This work is planned for the future.

4 Performance Evaluation

MoCCA is a GSM based cloud whose performance we evaluate here but it can act as a reference when cloud architectures based on 3G and 4G networks that support application bandwidths in excess of 20 Mbps are evaluated. With higher bandwidths such systems should perform better than GSM for cloud applications.

We have computed the performance of a mobile cellular cloud under the following scenarios consisting of different workloads, configurations, probability of blocking, probability of handoffs under different vehicular speeds, and energy consumptions.

Workloads: We have mainly chosen compute bound applications, such as: (1) Common matrix computations e.g. inversion, eigenvalues and eigenvectors, determinant, Fast Fourier transform, and Cholesky decomposition. (2) Sorting of large arrays and linear regression. (3) Fibonacci number calculations, etc.

Configurations: Same computations are performed under three configurations: (i) a single mobile station (no redundancy), (b) Double modular redundancy, (c) Triple modular redundancy.

Probability of blocking: Computations have been repeated for different probabilities.

Mobility: We have also computed resultswith different handoff rates, the probability of handoffs, and probability of handoff droppings under different mobility models.

Energy: For all these scenarios, we have computed the energy consumption for transmission, reception and computations to find the total energy expenditure related to cloud computation. However, due to space limitation, we are not able to include all the results in this paper.

For the analysis purpose, we consider one GSM cell that has 53 channels for uplink and 53 channels for downlink transmissions. A GSM frame consists of 8 time slots each of which has duration of 577 micro-seconds in which 114 bits of data can be carried. We assume a MS to be equipped with a 500 MHz processor with 256 Mbytes of memory that runs a MS windows operating system. Our main reason for choosing these parameters is that several compute bound applications running on similar hardware have been bench-marked [22] and quite a few commercial mobile phones available on the market have specifications within the range.

In the analysis that we present here, we have assumed that a base station keeps a fraction of channels for incoming voice traffic and the remaining for cloud applications. In order to compute the capacity of the cloud for the compute bound applications, we make use of the trunking theory [17]. We define λ to be the average number of session request rate and μ the service rate for each MS which gives us a traffic intensity of $A_u = \lambda\mu$ Erlangs, with the probability of blocking, p_m:

$$p_m = \frac{\frac{1}{m!}\left(\frac{\lambda}{\mu}\right)^m}{1 + \sum_{n=1}^{m}\frac{1}{n!}\left(\frac{\lambda}{\mu}\right)^n} . \tag{3}$$

Here m is the number of channels and n is the number of MSs in the coverage area. In the mobile cloud architecture the originator of the call is the

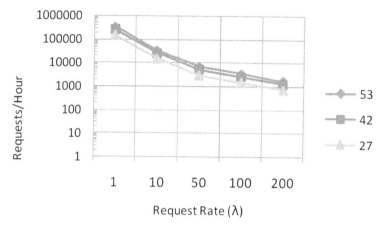

Figure 4 Cholesky decomposition.

base station and the call is terminated on mobile phones serving the agreed upon application. In the next sections we describe the cloud performance for several applications computed assuming no redundancy in the hardware.

4.1 Application 1 – Cholesky Decomposition

We assume a 100 by 100 matrix to be factorized on the mobile devices. Figure 4 plots the total number of such requests completed by the MSs when they utilize 53 channels (Blue Curve), 42 channels (Red Curve), or 27 channels (Green Curve) with respect to the request rates assigned to each MS by the base station. For example with 53 channels, with probability of blocking of 0.005, a maximum number of Cholesky decomposition that can be completed in one hour are 343,583 on participating mobile devices. In Figure 3, the x-axis depicts request rates assigned to mobile devices per hour by the base station. For example, when the base station assigns 1 request/hr to mobile devices using 27 channels, the maximum number of requests completed is 147000 per hour. But this number changes to 14700 per hour when each participating device is completing 10 requests per hour giving us the total request completions of $(10 \times 14700 = 147000)$ which is still the same.

So in this case the number of participating devices becomes 14700. The number of participating devices reduces to 1470 when each device is able to complete 100 requests per hour. So depending upon the number of participating devices and the number of channels, the base station can distribute the workload according to application needs.

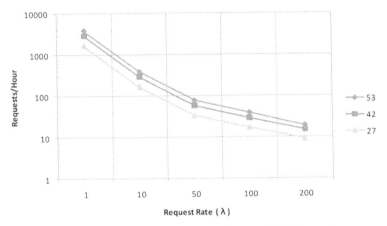

Figure 5 FFT computation – one iteration on 900,000 samples.

4.2 Application 2 – Fast Fourier Transform using S

The second compute bound application is related to the one iteration of Fast Fourier Transform (FFT) computation of 900, 000 data samples using the S [23] program package. We assume that S is running on the mobile. We use the same set of channels for this experiment. Figure 4 depicts the request completion rate. With 42 channels, 2905 instances of the applications can be run in an hour with $p_m = 0.005$.

This application requires a channel holding time of 36.5 seconds.

4.3 Application I Capacity

Figure 6 shows the capacity of application 1 with the probability of blocking for different number of channels which increases with the increase in the probability of blocking (P_m).

4.4 Handoff Performance

For the computation of the handoff probability, we make an assumption that the cloud session originated in the cell with $1/\mu$ as the duration of the session which is represented by a random variable T_n that has exponential distribution given by $f_n(t) = \text{Prob}(T_n \le t) = 1 - \mu e^{\mu t}$. We assume the mobile is moving within the cell with a constant velocity of V Kilo-meter/hr in a cell that has a radius r.

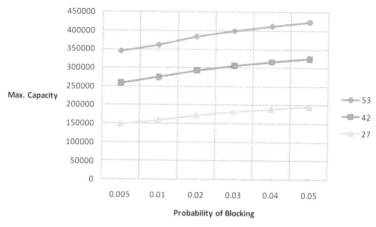

Figure 6 Probability of blocking.

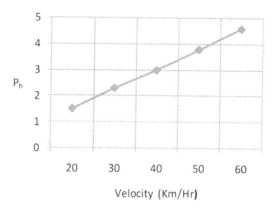

Velocity (Km/Hr)

Figure 7 Handoff probability.

Assuming a circular cell for simplicity, we can write the rate of cell crossing $\eta = 2V/\pi r$ which implies that mobile is dwelling in the cell for $1/\eta$ duration. Assuming cell dwell time to be exponentially distributed, it can be represented by the probability density function $f_h = \eta e^{-\eta T_h}$, where T_h is the random variable representing the dwell time. With these parameters, we can write the expression for the probability of handoffs:

$$P_h = \mathrm{Prob}(T_h > T_h) = \int_0^\infty \eta e^{-\eta T_h} \left[\int_{T_h}^\infty \mu e^{-\mu t} dt \right] dT_h = \frac{\eta}{(\eta + \mu)}. \quad (4)$$

Figure 7 depicts the variation of handoff probability with respect to the velocity of the mobile for a cell radius of 10 Kilometers for application 2. The probability of handoff varies from 1.5% at velocity of 20 Km/hr to 4.5% with velocity of 60 Km/hr. The handoff probability results suggest that for a typical cloud application the probability of handoffs is quite low at modest vehicular speeds.

4.5 Energy

A mobile phone participating in a cloud computing application expends energy in data computation, reception and transmission. Even after emergence of hardware and software features that help conserve the energy expenditure such as (a) turning the clock rate down when display is off, (b) dynamic voltage scaling, and (c) dynamic frequency scaling etc. saving power in mobile phones is still remains a major challenge. Recent power measurements that have been made on a mobile phone running numerical linear algebraic algorithms which are part of LINPACK benchmark [25] lead to the following observations and conclusions: (a) The energy expenditure per byte reduces with increasing packet sizes. For example a 64 byte packet may consume 31.25 milli-watts (mW) per byte when compared with a 512 byte packet consuming 3.906 mW per byte. (b) The energy expenditure per byte reduces when multiple packets are transmitted at the same time for example transmissions of 2, 8, 16, or 128 packets per second. The corresponding power consumption varies from 31.25 mW per byte to 0.49 mW per byte.

As stated earlier in the paper that the energy cost of computations is several order of magnitude less than the cost of transmissions or receptions, we only include energy results for the transmission and the reception for a few applications considered in the paper. Figure 8 depicts the variation of energy consumption with respect to packet size when only one packet per second is transmitted. The energy expenditure in Joules (J) varies from 770 J for a 64 byte long packet to 48.20 J for a 1024 byte packet when a Cholesky decomposition is performed on a 1000 × 1000 symmetric positive definite matrix Figure 6 shows the results for the energy consumption for application I with respect to number of packet size when different number of packets per second (pps) are transmitted. As one can notice that the energy consumption for a 128 byte packet is 12 J when 64 pps are transmitted and it reduces to 6.9 J when 128 pps are transmitted. The percent reduction in energy consumption is more than 50% which is significant. Figure 9 shows the energy expenditure in data reception with respect to packet size for Cholesky decomposition.

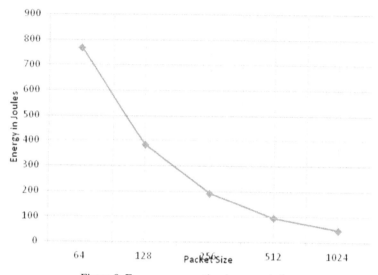

Figure 8 Energy consumption in transmission.

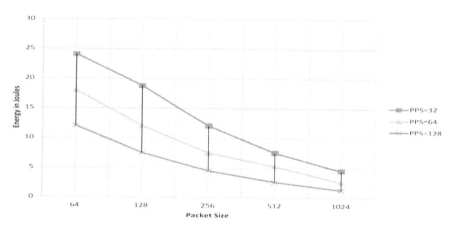

Figure 9 Energy consumption in transmission.

Using a 64 byte packet size the total energy expended in reception is 270 J when 2 pps are received and it reduces to 5.18 J for a 512 byte packet with a reception of 16 pps.

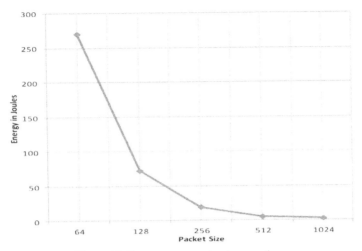

Figure 10 Energy consumption in receptiopn.

5 Related Work

In this section, we review the current research in the mobile cloud computing area. Ideas which appear similar to MoCCA have been proposed for sensor networks [25] and mobile grids [14]. But MoCCA is very different from sensor network based grids, mobile grids, or any other peer-to-peer network based grids. The important differences are: (1) Each MS belongs to a different owner, (2) Each MS is battery powered, (3) MSs are mobile, and therefore the nodes forming a cloud keep changing with respect to time, and as a consequence have (4) Dynamic cloud topology. Since the mobile devices do not directly communicate with each other i.e. there is no peer-to-peer communication, the impact of dynamic topology is not an issue in MoCCA, but this can be a major issue in clouds that rely on peer-to-peer networking. Mobile grid [1, 4, 11, 19] studies different aspects of mobile grid computing such as problem of multiple connects and disconnects, collaboration among mobile devices that are heterogeneous in terms of resources, interworking among mobile and stationery computers, integration of mobile phones to infrastructure based wireless grid. Both the wireless grid and mobile grid advocate some form of peer-to-peer networking.

WIPDroid [3] is a platform created for Droid phones [9] by integrating WIP (Web Service Initiation Protocol) for real-time service oriented communication over IP which has been mentioned as it could provide SaaS type of service as a client but not as a server which is the focus of this paper. Wireless sensor networks can constitute a valid approach to mobile cloud computing and in some ways this approach may resemble MoCCA, if we assume the sink of a sensor network to be acting like a base station of cellular networks, and the sensor field resembles the coverage area of the base station. But sensor networks and cellular networks have many profound differences e.g. the data transmission among sensor nodes is multi-hop whereas in cellular networks its only one hop. Finally, somewhat related to MoCCA are mobile data sharing systems which use distributed file systems and peer-to-peer systems such as [6, 10, 15] and mobile computing platforms such as Hyrax [11].

6 Conclusion

This paper presented MoCCA – an architecture for building mobile clouds that leverages cellular infrastructure, thus making it inexpensive and easy to deploy. The performance analysis of single GSM base station suggests that several numerical and statistical algorithms can run on large data sets in the cellular environment even with a small number of mobile phones participating only for several minutes. With the deployment of LTE that provides larger bandwidths and uses resource rich smart phones running at 1 GHz speed, mobile clouds have potential to become popular. As part of the future work, we are working on characterizing the overhead of MapReduce, additional control signaling, and energy consumption in transmission and computation. We are planning on to extend the current GSM model to a 4G LTE system that has much larger bandwidth and several attractive features for data applications.

References

[1] S. Ahuja and J. Myers. A survey on wireless grid computing. The Journal of Supercomputing, 37:1, 2006.

[2] Z. Chen, Energy-efficient Information Collection and Dissemination in Wireless Sensor Networks. PhD thesis, University of Michigan, 2009.

[3] W. Chou and L. Li. WIPdroid – A two-way web services and real-time communication enabled mobile computing platform for distributed services computing. In IEEE International Conference on Services Computing, Vol. 2, pp. 205–212. IEEE, 2008.

[4] D. Chu and M. Humphrey. Mobile OGSI. NET: Grid Computing on Mobile Devices. In 5th IEEE/ACM International Workshop on Grid Computing, pp. 182–191. IEEE Computer Society, 2004.

[5] J. Dean and S. Ghemawat. MapReduce: Simplified data processing on large clusters. Communications of the ACM, 51(1):107–113, 2008.

[6] G. Ding and B. Bhargava. Peer-to-peer file-sharing over mobile ad hoc networks. In Proceedings Second IEEE Annual Conference on Pervasive Computing and Communications, pp. 104–108. IEEE, 2004.

[7] Gartner Inc. Competitive Landscape: Mobile Devices, Worldwide, 1Q10. Online at http://www.gartner.com, May 2010.

[8] International Telecommunication Union, ITU Corporate Annual Report. Online at http://bit.ly/aJSfQs, 2009.

[9] O. Kharif. A warm welcome for Android. BusinessWeek, January 2008.

[10] C. Lindemann and O. Waldhorst. A distributed search service for peer-to-peer file sharing in mobile applications. In Proceedings 2nd IEEE International Conference on Peer-to-Peer Computing, pp. 73–80, 2002.

[11] E. Marinelli. Hyrax: Cloud Computing in Mobile Devices using MapReduce. Master's thesis, Carnegie Mellon University, 2009.

[12] J. McKenderick. The $80 data center: Cheap computing or head in the cloud? Online at http://zd.net.bBooxt, November 2007.

[13] L. McKnight, J. Howison, and S. Bradner. Wireless grids – Distributed resource sharing by mobile, nomadic, and fixed devices. IEEE Internet Computing 8(4):24–31, 2004.

[14] P. Mell and T. Grance. The NIST Definition of Cloud Computing. Version 15, 10-7-09, National Institute of Standards and Technology, 2009.

[15] N. Michalakis and D. Kalofonos. Designing an NFS-based mobile distributed file system for ephemeral sharing in proximity networks. In Proceedings 4th Workshop on Applications and Services in Wireless Networks (ASWN), pp. 225–231. IEEE, 2005.

[16] R. Miller. Microsoft's Windows Azure Cloud Container. Online at http://bit.ly/cenSFw, November 2009.

[17] M. Mouly and M. Pautet. The GSM system for mobile communications. Telecom Publishing, 1992.

[18] J. Neumann. Probabilistic logics and the synthesis of reliable organisms from unreliable components. Automata studies (1956), 43–98.

[19] N. Palmer, R. Kemp, T. Kielmann, and H. Bal. Ibis for mobility: Solving challenges of mobile computing using grid techniques. In Proceedings 10th Workshop on Mobile Computing Systems and Applications, p. 17. ACM, 2009.

[20] M. Patterson, D. Costello, P. Grimm, and M. Loeffler. Data center TCO. A comparison of high-density and low-density spaces. Thermes, 2007.

[21] A. Mishra. Security and Quality of Service in Ad hoc Wireless Networks. Cambridge University Press, 2008.

[22] Stephen Stenhouse's benchmark, version 2, www.sciviews.org/benchmark/benchmark1.htm

[23] Richard Becker et al. The New S Language. Wadsworth & Brooks/Cole, 1988.

[24] Erik Dahlman et al. 4GLTE/LTE Advanced for Mobile Broadband. Academic Press, 2011.

Biographies

Amitabh Mishra is a faculty in the Information Security Institute of Johns Hopkins University in Baltimore, Maryland. His current research is in the area of cloud computing, data analytics, dynamic spectrum management, and data network security and forensics. In the past he has worked on the cross-layer design optimization of sensor networking protocols, media access control algorithms for cellular-ad hoc inter-working, systems for critical infrastructure protection, and intrusion detection in mobile ad hoc networks. His research has been sponsored by NSA, DARPA, NSF, NASA, Raytheon, BAE, APL, and US Army. In the past, he was associate professor of computer engineering at Virginia Tech and a member of technical staff with Lucent Technologies – Bell Laboratories in Naperville, Illinois. His has worked on architecture and performance of communication applications running on 5ESS switch. GPRS, CDMA2000 and UMTS were a few of the areas he worked on while with Bell Laboratories. He received his B. Eng. and M. Tech. degrees in Electrical Engineering from Government Engineering College, Jabalpur and Indian Institute of Technology, Kharagpur in 1973, and 1975 respectively. He obtained his M. Eng. and Ph. D. in 1982, and 1985 respectively also in Electrical Engineering from McGill University, and a MS in Computer Science in 1996 from the University of Illinois at Urbana-Champaign. Dr. Mishra is a senior member of IEEE, a member of ACM, and SIAM. He is author of the book *Security and Quality of Service in Wireless Ad hoc Networks*, published by Cambridge University Press (2007). He is a technical editor of *IEEE Communications Magazine*.

Gerald Masson is a Professor (Emeritus) of Computer Science at Johns Hopkins University, Baltimore, Maryland. He was the founding director of Johns Hopkins University Information Security Institute and chair of the computer science department. His research interests are in fault tolerant computing, real-time error monitoring of hardware and software, inter-connection networks, and computer-communications. He is a Fellow of IEEE.

Ivy: Interest-based Data Delivery in VANET through Neighbor Caching

Tan Yan and Guiling Wang

Department of Computer Science, New Jersey Institute of Technology, University Heights, Newark, NJ 07102, USA; e-mail: {ty7,gwang}@njit.edu

Received 20 May 2013; Accepted 15 July 2013

Abstract

In this paper, we study the problem of interest-based data delivery in Vehicular Ad Hoc Networks (VANETs), which is to efficiently forward data to a vehicle that owns the given interest without knowing its ID beforehand. Such problem is generally challenging in large-scale distributed networks, because it usually requires to query a huge number of nodes in the network to find a node of interest out of them, which is costly. To tackle the problem, we design an interest-based data delivery (*Ivy*) scheme through neighbor caching to boost the efficiency in querying information of the vehicle with given interest. We let each vehicle cache the information of all the neighbor vehicles it meets during driving, such that by just querying a few vehicles, a vehicle can retrieve the information of a large amount of vehicles, which greatly reduces the message overhead. To further reduce the message cost in route establishment, we calculate an estimated current location of the interested vehicle and forward data towards the calculated location, which avoids broadcasting blindly to search for the vehicle. Simulation result shows *Ivy* is both efficient and effective, and outperforms existing data delivery schemes with higher delivery ratio, lower delay and smaller message overhead.

Keywords: Interest-based, data delivery, neighbor caching.

Journal of Cyber Security and Mobility, Vol. 2, 127–149.

1 Introduction

A Vehicular Ad hoc Network (VANET) is an on-road network where vehicles exchange and deliver data through multi-hop communication. Data delivery in VANET is envisioned to be applied into many promising applications for both safety [23] and non-safety [19] proposes. To improve the delivery efficiency, quite a few routing and data dissemination schemes have been designed to effectively forward data to vehicles with known identity [26] or in a given location/area [35, 20].

However, the current data delivery building blocks are not sufficient to support many applications in VANETs, which generally require to deliver data to vehicles based on their interests. For example, in data dissemination services [32], a disseminator wants to find a vehicle interested in such services (e.g., subscribes to the services) to relay the dissemination. In a rural area far from hospitals, a driver in a crashed car wants to find a vehicle interested in providing on-road rescue (e.g., the driver is a doctor or a mechanical technician) and ask for help. We call such applications, *interest-based* data delivery, in which, the data source needs to deliver data to a vehicle that has the given interest, and does not know its ID or location beforehand. Traditional data forwarding mechanisms fail to effectively support interest-based data delivery, since they require the source to have a priori knowledge of either the ID or the location of the destination. Moreover, even the source knows the ID and does not know the location of the destination, it still has to broadcast intensively to all the directions to build a route with destination, which is very costly.

In addressing the problem of a source delivering data to a destination with given interest, the primary objective is to reach the destination with small overhead and low delay. A solution direction could be first querying and obtaining the ID and the current location of the interested vehicle, and then making the source only forward data towards the obtained location to reach the vehicle, which avoids broadcasting blindly to the network and greatly reduces the message overhead. However, the costly query procedure brings us a challenge in making the solution feasible in distributed networks, especially when the network has a large number of vehicles and only a few of them are with the given interest. The source vehicle has to propagate the query message to almost the entire network in order to find the information of an interested vehicle. In addition, as vehicles are of high mobility, obtaining an accurate location of an interested vehicle is even more challenge. For example, to get the most up-to-date location of a vehicle, the only way is to ping this vehicle and obtain its current location, which simply brings us back to the original problem: how to reach this vehicle? Therefore, in reaching a vehicle with given interest, an efficient method that obtains the ID and the current location of the interested vehicle is needed.

This paper focuses on designing a distributed interest-based data delivery protocol, which delivers data to a vehicle that has given interest with small message overhead and low delay. To be more specific, our goal in this paper is to solve the following problem: *In a network where each vehicle has a collection of interests, given a specific interest, how to efficiently identify and reach a vehicle that owns the given interest without knowing its ID or location beforehand?*

We propose *interest-based data delivery (Ivy)* to deliver data to a vehicle with given interest through neighbor caching. *Ivy* first efficiently queries and obtains the information, such as the ID and the current location, of the vehicle with given interest, and then uses such information to establish a route to reach the vehicle with conserved message overhead. To facilitate vehicle information query, we let each vehicle include its information into redesigned beacon messages, through which, neighbor vehicles can exchange and store the information for each other. To boost the query efficiency, we design a Cached Neighbor List for each vehicle to cache the beacon information received from all the vehicles it meets during driving, so that each vehicle in the network can store information of many other vehicles. In this way, we can retrieve the information of a large amount of vehicles by just querying a few of them, which greatly reduces the message overhead in obtaining the information of the interested vehicle. To reduce the message cost in route establishment, we estimate the current location of the interested vehicle by analyzing the location, the speed and the driving direction in the cached information, and we only forward data towards the estimated location to build a route with the interested vehicle. To cope with the vehicle mobility, we keep querying and updating the information of the interested vehicle during data forwarding, and use such information to dynamically adjust the estimated location and the data forwarding direction. Simulation result shows *Ivy* is both efficient and effective in reaching a vehicle with given interest, and outperforms existing data delivery schemes with higher delivery ratio, lower delay and smaller message overhead.

The rest of this paper is organized as follows. We introduce the assumption and the overview of the *Ivy* in Section 2. The detailed protocol of *Ivy* is presented in Section 3. Section 4 describes the enhanced scheme with location refinement. An evaluation of *Ivy* is given in Section 5. Section 6 presents the related work. Finally, we conclude the paper in Section 7.

2 Assumptions and Scheme Overview

2.1 Assumptions

We assume vehicles communicate with each other through a dedicated short-range wireless channel (DSRC [1]) (100 to 250 m) in VANET. Each vehicle is preloaded with a digital street map, and knows its location that can be obtained through GPS device or various on-road localization mechanisms [33]. Vehicles can find their neighbors through periodical exchange of beacon messages, which can be done efficiently with various of protocols such as S-Aloha [12, 13].

We assume each vehicle tags itself with labels of interests showing the services it is interested in or the applications it subscribes, which is widely used in many publish/subscribe and social systems [28]. Vehicles discloses their interests to the network in order to provide/receive the services they subscribe.

2.2 Overview of *Ivy*

Ivy is a distributed protocol in VANET to make the source vehicle efficiently identify and obtain the information of a vehicle that has the given interest, estimate the current location of the vehicle, and deliver data to it with conserved message overhead. As the network may only have a small number of vehicles with given interest and they are randomly distributed in the network, to obtain their information, literally we have to search and check the information of a large number of vehicles before finding the information of an interested vehicle. Moreover, since each vehicle in the network only has the knowledge of limited number of vehicles in the network, we need to send messages to query multiple vehicles in order to hit the one that has the information of the interested vehicle. Thus, to reduce the message overhead in obtaining the information of an interested vehicle, we need to make each vehicle store the information of as many vehicles as possible, so that we can retrieve the information of a large number of vehicles by only querying a few of them.

We observe that a vehicle driving around an area meets different vehicles. Through beacon message exchange, the vehicle can obtain the information of the vehicles it meets. The more vehicles it meets, the more vehicle information it can possibly obtain. Inspired by this, we let each vehicle embed its information, such as ID, interests, location and driving direction, into the beacon messages and exchange it with neighbors. We design a Cached Neighbor List for each vehicle to cache the beacon messages received from neighbor vehicles it meets when driving around the area, which stores the information of each neighbor vehicle at the time they meet. When a source needs to reach a vehicle with given interest, it broadcasts query messages looking for information of the vehicle that has the given interest. Upon receiving the message, each vehicle searches in its Cached Neighbor List, and replies to the source if it finds any vehicle whose interests contains the given interest. After collecting all the reply messages, the source obtains the ID of the interested vehicle, as well as the location and the driving direction that are recorded at the time the vehicle meets the reply sender. To calculate the current location of the interested vehicle, we first analyze the traffic of the area and calculate a speed range of the vehicle. Then we compute a region, which is a collection of all the locations the interested vehicle can reach under the calculated speed range. The calculated region is used as the estimation of current location of the interested vehicle. We further propose several heuristics to refine the region, such as by applying driving direction of the vehicle and by aggregating information from multiple reply messages. Finally, we route the message towards the estimated region and broadcast in the region to reach the vehicle.

As vehicles are highly mobile, it is possible that when the message reaches the estimated region, the interested vehicle has already moved out of the region and cannot receive the broadcasted message. To mitigate this issue, we enhance *Ivy* with location refinement (*Ivy-LR*), which dynamically adjusts the estimated region and the data forwarding direction by keeping querying the information of the interested vehicle when forwarding data.

In the following, we present the detailed design of *Ivy* and *Ivy-LR* in Sections 3 and 4, respectively.

3 Ivy: Interest-based Data Delivery through Neighbor Caching

In this section, we present Interest-based Data Delivery (*Ivy*) through Neighbor Caching. We first introduce the structure design of the beacon message and the proposed Cached Neighbor List, through which vehicles exchange their information and cache the received neighbor information. Then, we present interest-based information query for the source vehicle to efficiently

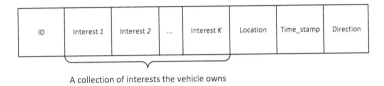

A collection of interests the vehicle owns

Figure 1 Structure of the beacon message.

obtain the information of a vehicle that has the given interest. After that, we design location estimation to calculate the current location of the interested vehicle based on the obtained information. Finally, we propose data delivery to the estimated location to reach the vehicle.

3.1 Structure of Beacon Message and Cached Neighbor List

Vehicles obtain their neighbor information through the exchange of beacon messages. To facilitate querying the vehicle with given interest, we redesign the structure of the beacon message by adding the *ID*, *Interests*, *Time-stamp*, *Location*, and *Driving direction* fields to it. Figure 1 shows the structure of the beacon message. The detailed description of each field is shown as follows:

- *ID*: the ID of the vehicle that sends the beacon message.
- *Interests*: the list of interests the vehicle wants to disclose to public, which can contain multiple elements. For example, a driver may want to add 'Game' and 'Spanish' into this field to subscribe to game and Spanish services/channels.
- *Time_stamp*: the time when the beacon message is sent out.
- *Location*: the position of the vehicle when the beacon message is sent out.
- *Direction*: the driving direction of the vehicle, i.e., North, South, East or West.

During driving, each vehicle periodically broadcasts the beacon messages and exchanges with its neighbor vehicles.

We design a Cached Neighbor List for each vehicle to parse the received beacon messages and store the information of the neighbor vehicles it meets during driving. The Cached Neighbor List is a queue, which is divided into two zones according to the time the vehicle receives the information: (1) *Current Neighbor Zone* that contains the information of the recently discovered neighbor vehicles, considering these vehicles may still be its neighbors, and (2) *Historical Neighbor Zone* that contains the information of the neighbors

	ID	Interests	Location	Time_stamp	Direction
Current Neighbor Zone Information received less than t_c seconds ago	ID_1	$I_{1a} \dots I_{1k}$	(x_1, y_1)	t_1	North
	:	:	:	:	:
	ID_m	$I_{ma} \dots I_{mk}$	(x_m, y_m)	t_m	South
Historical Neighbor Zone Information received more than t_c but less than t_e seconds ago	:	:	:	:	:
	ID_n	$I_{na} \dots I_{nk}$	(x_n, y_n)	t_n	East

Figure 2 Structure of the Cached Neighbor List.

that are discovered in the past, and these vehicle are no longer its neighbors. Figure 2 shows the structure of the Cached Neighbor List, where t_c and t_e are system parameters.

Basically, each vehicle caches its neighbor information for up to t_c seconds. Each time when a vehicle receives a new beacon message, the vehicle first parses and stores the information into the Current Neighbor Zone of its Cached Neighbor List. Then, the vehicle moves all the information that was received t_c seconds ago into the Historical Neighbor Zone. After this, it deletes the information that was received t_e seconds ago.

3.2 Interest-based Information Query

When a source wants to obtain the information of a vehicle that has the given interest, it broadcasts query messages to nearby vehicles to check whether they have the information of the interested vehicle in their Cached Neighbor List. A vehicle replies to the source vehicle with such information if it finds any.

In detail, when the query procedure starts, the source vehicle first specifies the interest it needs, and encloses the interest information into the head of the query message. After that, the source sets the Time-to-Live (TTL) of the query message and broadcasts to its neighbors. Upon receiving the query message, each vehicle searches its Cached Neighbor List to see whether it contains the information of a vehicle with the given interest. The vehicle first checks the Current Neighbor Zone. If it finds an entry whose *Interests* field contains the given interest, it collects the vehicle ID from the *ID* field of this entry, and then simply unicasts one-hop to reach the vehicle, considering they are currently neighbors. Upon receiving the message, the vehicle with the given interest replies to the source vehicle to establish a communication route. Up on receiving the query message, if a vehicle cannot find any information containing the given interest in the Current Neighbor Zone, the vehicle further checks its Historical Neighbor Zone. If the Historical Neighbor Zone contains such entries, the vehicle replies to the source with the entries it finds. After completing the search in its Cached Neighbor List, the vehicle reduces the TTL value of the message by one and further propagates the query message to its neighbors. A query message is dropped if its TTL value reaches zero.

After the source vehicle sends out the query messages, it sets a timer waiting for the reply. Upon timer firing, if the source cannot receive any reply, it resets the timer, increases the initial TTL value of the message by two, and rebroadcasts the message.

The detailed procedure of the Interest-based Information Query is illustrated in Algorithm 1.

3.3 Location Estimation

After receiving all the reply messages, the source vehicle parses the information from the messages, based on which, it obtains the ID and calculates an estimated current location of the interested vehicle. More specifically, from each reply message, the source obtains the ID of the vehicle that owns the given interest, and the location, driving speed and driving direction of this vehicle at the time the vehicle meets the sender of the reply message (i.e., value of *Location*, *Speed*, and *Direction* field at the time in *Time_stamp* field). Since vehicles are mobile, the current location of the interested vehicle is not the same as that in the reply message, and the speed of the vehicle may also vary over the time. To calculate the current position of the vehicle, we first estimate the minimum speed v_{min} and maximum speed v_{max} of the vehicle. Then we compute a region, a collection of all the locations the vehicle can reach under a speed within $[v_{min}, v_{max}]$. We further refine the calculated region by applying vehicle's driving direction and aggregating information from multiple reply messages. The refined region is the estimated current position of the vehicle.

3.3.1 Speed Estimation

To calculate the speed range of the interested vehicle, we adopt the widely applied Gaussian Distribution to model vehicle speed [34]. The speed v of a

---

Algorithm 1 The Interest-based Information Query

Notations:

Interest_given: the given interest specified by the source node.
i.X: the field X of entry i in a Cached Neighbor List.
Qualified_List: a list of entries that contain information of qualified vehicles.

At the source vehicle
1: Set the TTL value;
2: Set a timer;
3: Broadcast the query message;
    Upon the timer fires
4: **if** receive no reply **then**
5:      TTL=TTL+2;
6:      **GOTO** STEP 2
7: **else**
8:      **Exit**;
9: **end if**
   At the vehicle whose interests contains the given interest
   Upon receiving the query message
10: Send reply to the source node;
   At the vehicle whose interests does not contains the given interest
   Upon receiving the query message
11: **for** i in Current Neighbor Zone **do**
12:     **if** *Interest_given* is in *i.Interest* **then**
13:         Forward the query message to vehicle with *i.ID*;
14:         **Exit**;
15:     **end if**
16: **end for**
17: **for** i in Historical Neighbor Zone **do**
18:     **if** *Interest_given* $\in$ *i.Interest* **then**
19:         Add entry i to *Qualified_List*;
20:     **end if**
21: **end for**
22: **if** *Qualified_List* != $\emptyset$ **then**
23:     Reply *Qualified_List* to source vehicle;
24: **end if**
25: **if** $--TTL$!=0 **then**
26:     Broadcast the query message;
27: **end if**

Table 1 Confidence interval under different probabilities.

Probability (p)	Confidence Interval/Speed Range
0.8	$\mu \pm 1.281\sigma$
0.9	$\mu \pm 1.645\sigma$
0.95	$\mu \pm 1.960\sigma$
0.98	$\mu \pm 2.326\sigma$
0.99	$\mu \pm 2.575\sigma$
0.995	$\mu \pm 2.807\sigma$

vehicle can be modeled as follows:

$$v \sim N(\mu, \sigma^2), \tag{1}$$

where μ is the mean and σ is the standard deviation. Given an area, the mean and standard deviation of the vehicle speed in the area can be obtained through various traffic statistical services, such as Google Maps [2] and Microsoft Bing Maps [3].

Given the vehicle speed in an area modeled with Gaussian distribution shown as in Eq. (1), we want to compute a speed range [v_{min}, v_{max}], which is a confidence interval, such that the speed of a vehicle has a high probability (e.g., 0.9) to fall in this interval. We use quantile function to calculate the confidence interval of the vehicle speed under a given probability. The quantile function of a distribution is the inverse of the cumulative distribution function. For a Gaussian distribution with mean μ and standard deviation σ, given a probability p, the quantile function of the distribution can be written as:

$$F^{-1}p = \mu + \sigma\sqrt{2} \cdot \mathrm{erf}^{-1}(2p - 1), \tag{2}$$

where *erf* is the error function shown as follows:

$$\mathrm{erf}(x) = 2/\sqrt{\pi} \cdot \int_0^x e^{-t^2} dt. \tag{3}$$

Table 1 shows the calculated confidence interval (speed range) under different probabilities.

For an interested vehicle, we specify a high probability value (e.g., 0.9 or 0.95), and then calculate a speed range of this vehicle to ensure the actual speed of the vehicle has a high probability to fall in the calculated range. For example, in an area, when the mean of the vehicle speed is 30 m/s and the standard deviation is 5m/s , given the probably of 0.9, the corrsponding speed range of a vehicle in the area is $30 \pm 1.645 \times 5 = [21.775$ m/s, 38.225 m/s$]$.

3.3.2 Calculation of the Possible Locations

We compute the speed range of the interested vehicle. In addition, from each received reply message, we can obtain the time the interested vehicle meets the reply sender (value in *Time-stamp* field) and the location at the time it meets the reply sender (value in *Location* field). By comparing the time recorded in the reply message and the current time, we can further calculate the time duration the vehicle travels after meeting with the reply sender. Combining all such obtained information, we can estimate the current region of the interested vehicle, which is a collection of all the possible locations this vehicle can reach after meeting with the reply sender, under the calculated speed range.

In detail, when the source vehicle receives a reply message from the reply sender containing the information of an interested vehicle, it first computes the speed range $[v_{min}, v_{max}]$ of the vehicle using the method introduced in Section 3.3.1. Then, it calculates the traveling time t_{tr} of the vehicle by computing the time difference between the value in *Time_stamp* field of the reply message and the current time. t_{tr} is the time the vehicle travels after exchanging a beacon message with the reply sender. After that, the traveling distance d_{tr} of the vehicle can be computed by multiplying the speed of the vehicle and the calculated traveling time, which is the range shown as follows:

$$[v_{min} \times t_{tr}, v_{max} \times t_{tr}]. \qquad (4)$$

Thus, the possible current location of the interested vehicle is a region, a collection of locations whose distance to the location the vehicle meets the reply sender falls in the range shown in Eq. (4). For example, given the location where the interested vehicle meets the reply sender as shown in Figure 3, the estimated region is the area of a circle with the meeting location as the origin and $v_{max} \times t_{tr}$ as radius subtracting the area of a circle with the meeting location as the origin and $v_{min} \times t_{tr}$ as radius.

As vehicles are driving rationally, a vehicle may always drive on a road segment and does not go to the off-road area, and it may always try to drive in the direction towards the destination and will not drive back and forth through it can do so. For the interested vehicle, after departing from the location it meets the reply sender, it is very likely that the vehicle will keep driving in the direction same as the one when it meets the reply sender, i.e., the direction recorded in the *Direction* field of the reply message. Thus, the estimated region of the interested vehicle can be further refined, which is the collection of subregions that meet the following two conditions:

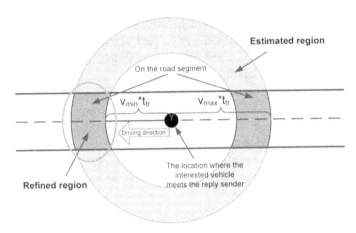

Figure 3 Region of the interested vehicle.

1. on a road segment;
2. in the driving direction (i.g., *Direction* field of the reply message) of the location where the interested vehicle meets the reply sender (e.g., *Location* field in the reply message).

For example, in Figure 3, since the vehicle drives to the left of the road, the area on the left-hand side of the figure is the refined region the interested vehicle.

As a vehicle driving in an area meets multiple vehicles, it is very likely that the information of the interested vehicle is cached by multiple vehicles, and the source vehicle may receive multiple reply messages containing the information of the same interested vehicle. We aggregate the information from multiple reply messages to provide more accurate location estimation for the interested vehicle. When the source receives multiple replies containing the information of the same vehicle, for each received message, the source applies the aforementioned method to calculate a region of possible locations for the vehicle. The aggregated region is the one that intersects all the calculated regions, considering the vehicle has a very high probability to present in this region. For example, a source vehicle receives three reply messages for the same vehicle. After applying the location estimation method, it calculates three sets of subregions $\{R_a, R_b, R_c, R_d\}$, $\{R_b, R_c, R_d\}$, and $\{R_a, R_b, R_c\}$. Thus, the final aggregated region of the vehicle is $\{R_b, R_c\}$.

It is also possible that the source vehicle will receive the information of multiple vehicles, each having the given interest. In this case, the source calculates the region for all these vehicles and picks the vehicle whose region is closest to the current location of the source as the interested vehicle.

3.4 Data Delivery

After *Interest-based Information Query* and *Location Estimation*, we obtain the ID and the estimated region of the vehicle that has the given interest. The estimated region is not the exact current location of the interested vehicle, but a set of subregions containing all the locations where the vehicle currently has a high probability to present. To deliver the data to the interested vehicle, we build a route from the source vehicle towards the calculated region and broadcast in the region to reach the interested vehicle. The routing and broadcasting can be carried out by various location-based data dissemination mechanisms, such as Geocast [11]. If the source vehicle cannot reach the interested vehicle after broadcasting, it doubles the size of the region and broadcasts the message to the expended region. If the source still cannot reach the destination, it cancels the broadcasting and starts sending the query messages to redo the whole data delivery procedure.

In Section 3.5 we will show an example of the whole procedure of the *Ivy* protocol.

3.5 An Example of the *Ivy* Protocol

Suppose the source vehicle A wants to find a vehicle whose interests includes *game* and deliver data to this vehicle as shown in Figure 4(a). Suppose vehicle B is a vehicle that has interest *game*. During driving, at time t_1, vehicle B meets vehicle C. Through beacon message exchange, C caches the information of B into its Cached Neighbor List. At time t_2, B meets another vehicle, D. Through beacon message exchange, D caches the information of B into its Cached Neighbor List. At time t_3, vehicle A broadcasts query message as shown in Figure 4(b). Upon receiving the query message, vehicle C and D reply to A with their cached information of vehicle B. Upon receiving reply messages, vehicle A obtains the ID of vehicle B and calculates an estimated region for each received message shown as the circle areas in Figure 4(b). According to the driving direction of vehicle B recorded in reply messages, vehicle A only considers the left half area of the circle with origin (x_1, y_1) and the lower half area of the circle with origin (x_2, y_2). After that, it computes the

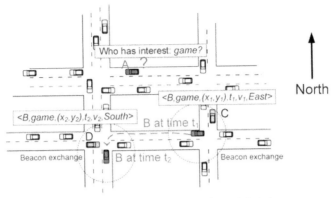

(a) Vehicle A needs to find a vehicle whose interests include *game*

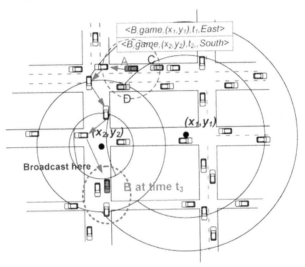

(b) At time t₃ (current time), vehicle A broadcasts query message and calculates the location of the interested vehicle, vehicle B

Figure 4   An example of the *Ivy* protocol.

aggregated region shown as the dashed oval in Figure 4(b), and then routes the data and broadcast it in the region.

4 *Ivy-LR*: Ivy with Location Refinement

As vehicles are of high mobility, after the source vehicle routes the data to the estimated region, it is possible that the interested vehicle may have already moved out of the region, which makes it fail to be reached by the source vehicle.

To mitigate this issue, in this section, we design an enhanced *Ivy* with location refinement (*Ivy-LR*) which dynamically adjusts the estimated region and the forwarding direction by keeping querying the information of the interested vehicle during data forwarding to receive more up-to-date information. The design of *Ivy-LR* is inspired by the following observation. For the information (e.g., Location and Driving direction) of a vehicle cached by another vehicle, the more recently the two vehicles meet, the more up-to-date the cached information is, and the more accurate the estimated region will be. Thus, the estimated region calculated using the information cached in vehicles that are close to the current location of the interested vehicle is more accurate than that using the information cached in vehicles far from that location. Therefore, when forwarding messages from the source vehicle to the estimated region, if we keep broadcasting queries in each hop we move closer to the region, we will receive more and more up-to-date locations and driving directions of the interested vehicle, which helps us update the estimated region and dynamically adjust the forwarding direction to reach the interested vehicle.

In detail, after the source vehicle calculates the region for the interested vehicle, it routes data to a vehicle closer to the calculated region. Upon receiving the message, the vehicle first broadcasts a query message to its neighbors asking for the information of the interested vehicle. Upon receiving the query message, the neighbor vehicle replies to the query sender if it has such information. Upon collecting the information from its neighbors, the vehicle computes an estimated region using the collected information, aggregates this region with the region computed by the source, and generates a refined region that is the area intersecting both regions. After that, the vehicle picks its neighbor closest to the newly generated region and forwards the message to it. The whole procedure terminates if the message reaches the interested vehicle.

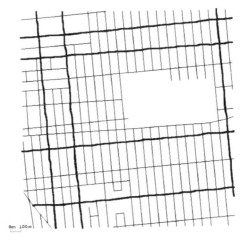

Figure 5 Simulation map.

5 Performance Evaluation

5.1 Evaluation Methodology

Ivy is evaluated through simulation implemented in the NS-2 simulator. The objective of the evaluation is three-fold:

1. Test the performance of *Ivy* in reaching a vehicle that has given interest.
2. Evaluate the efficiency of *Ivy* to reach the vehicle with given interest.
3. Test the effectiveness of location refinement in *Ivy-LR*.

We employ three metrics to do the evaluation. We choose the *delivery ratio* and *delivery delay* as measures of the performance and choose and *message overhead* as measures of cost. We compare our scheme with AODV [26] and GPSR [17], two widely used routing protocols in VANET and Mobile Ad Hoc Networks (MANETs). The original AODV and GPSR do not include interest information in their beacon messages. To have a fair comparison, we extend AODV and GPSR by adding interest information into the beacon message of each vehicle. In a network, there may be multiple vehicles that have the given interest, the source vehicle is considered to successfully deliver the data if it can reach any of the interested vehicle.

In our simulation, the map is a 2000 m × 2000 m area centered at (Latitude: 40.6548, longitude: –73.942795) in Brooklyn, NYC, extracted from the TIGER/Line (Topologically Intergraded Geographic Encoding and Referencing) database of the US Census Bureau [4]. Figure 5 shows the

Table 2 Simulation setting.

Parameter	Value
Size of the simulation area	2000 m × 2000 m
Simulation time	5000 s
Number of vehicles	200
Number of vehicles with given interest	5, 15
Vehicle communication range:	200 m
Vehicle speed	8 ~ 15 m/s

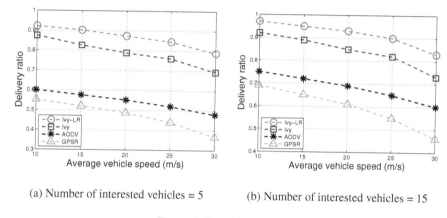

(a) Number of interested vehicles = 5   (b) Number of interested vehicles = 15

Figure 6 Data delivery ratio.

map we use. For the moving trace of vehicles, we employ the open-source, microscopic, space-continuous and time discrete vehicular traffic generator package SUMO [5] to generate the movements of vehicle nodes. SUMO uses a collision-free car-following model to determine the speeds and positions of the vehicles. After intergrading the SUMO trace into the map of TIGER, we discard the first 2000 seconds to obtain more accurate node movements. The output from SUMO is converted into input files for the movement of node in the NS-2 simulator. We configure the simulation according to the WAVE protocol [6]. Other system parameters are listed in Table 2.

5.2 Data Delivery Ratio

Figure 6 shows the data delivery ratio under different average vehicle speed. From the figure we can see that the delivery ratio of *Ivy-LR* is higher than that of *Ivy*, and they both outperform the compared schemes. When the average vehicle speed increases, the delivery ratio of all the schemes decreases. When

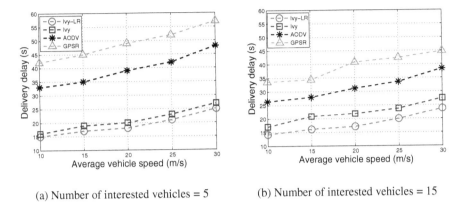

(a) Number of interested vehicles = 5   (b) Number of interested vehicles = 15

Figure 7   Delivery delay.

the number of interested vehicles in the network increase, the delivery ratio increases, which is because when the network has more vehicles with given interest, it is easier to identify and reach one out of them.

5.3 Data Delivery Delay

Figure 7 shows the data delivery delay under different average vehicle speed. From the figure we can see that the delivery delay of *Ivy-LR* is lower than that of *Ivy*, because *Ivy-LR* dynamically adjusts the estimated region to reach the destination, which improves the effectiveness of the data forwarding. Both of our schemes have lower delay than the compared schemes. When the average vehicle speed increases, the delivery delay of all the schemes increase because the forwarding link in the network may break more frequently, which increases the time in finding alternative routes. When the number of interested vehicles in the network increases, the delivery delay decreases, because the density of the interested vehicles increases, which reduces the routing length between the source vehicle and an interested vehicle.

5.4 Message Overhead

Figure 8 shows the number of transmitted messages under different average vehicle speed. From the figure we can see that the message overhead of *Ivy-LR* is higher than that of *Ivy*, because *Ivy-LR* needs to keep querying the information of the interested vehicle when forwarding data. Both *Ivy* and

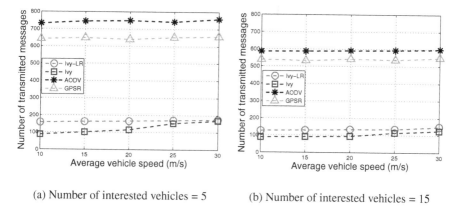

(a) Number of interested vehicles = 5 (b) Number of interested vehicles = 15

Figure 8 Message overhead.

Ivy-LR have less message overhead than the compared schemes, because our schemes only forward data to the direction towards the estimated region, while AODV and GPSR need to broadcast to all the directions to build a route to the destination. When the average vehicle speed increases, the message overhead of *Ivy* increases, while the overhead of the other three schemes does not increase significantly. In *Ivy*, if the speed of the interested vehicle is high, when the source routes the data to the estimated region, it is very likely that the vehicle has already moved out of the region. The source has to broadcast to a bigger region to search for the interested vehicle, which increases the message overhead. For *Ivy-LR*, since it keeps updating the estimated region, the message can always be forwarded to the location where the interested vehicle currently is. Since AODV and GPSR does not estimate the destination location, their message overhead does not change much when the vehicle speed changes.

6 Related Work

Data delivery has been extensively studied in both Mobile Ad Hoc Networks (MANETs) and Vehicular Ad Hoc Networks (VANETs). Node centric routing protocols, such as AODV [26], DSDV [27], DSR [16] and OLSR [9], establish topological end-to-end paths for the source to reach the destination. Their performance is further improved in [10, 24, 29] by exploiting the knowledge of vehicles' relative velocities and their constrained movements. Geographical routing protocols, such as GPSR [17], GOAFR [18] and GFG [7], leverage node positions to forward messages. To improve the greedy forwarding according to node's position, recovery mechanisms are provided in [22, 8], which proactively detect the potential dead-end positions and over-hears the communication channels to decrease the number of hops on the recovery paths. GSR [21] and SAR [30] adapts the concepts of anchor-based routing that was originally designed for sensor networks to vehicular network environments, in which a source vehicle uses a list of intersections to identify road segments and computes the shortest road-based path from its current position to the destination. A-STAR [50] improves GSR by giving preference to streets served by transit buses each time a new intersection is to be added to the source route. CAR [25] and GyTar [15] dynamically find connected paths between source and destination pairs by considering vehicular traffic and road topology. Opportunistic data forwarding mechanisms are designed in MDDV [31] and VADD [35]. VADD analyzes the historical traffic flow of an area to determine the best route to the destination. In addition to traffics, MDDV considers the number of lanes on each road segment to select a suitable road-based trajectory to forward data. In both protocols, when an in-termediate vehicle temporarily cannot reach another vehicle to forward data, a carry-and-forward approach is used, which caches the data packets until a suitable relay is found. Huang et al. [14] presents a delay tolerant epidemic routing protocol, which improves the delivery ratio in the scenario with very sparse vehicle traffic.

The aforementioned schemes together improves the performance and effi-ciency for delivering data to a vehicle with known ID or in a known location. However, none of them considers reaching a vehicle that has given interest without knowing its ID or location beforehand. Without a delegated mech-anism to efficiently obtain the ID and calculate the location of the interested vehicle, these mechanisms cannot be applied to achieve the goal of the paper.

7 Conclusion

This paper proposes *Ivy*, an interest-based data delivery scheme to deliver data to a vehicle that has a given interest. We increase the efficiency in querying and identifying the vehicle of interest by letting each vehicle in the network cache the neighbor vehicles it meets during driving, such that by querying only a few vehicles, we can retrieve the information of a large amount of vehicles and identify the interested vehicle out of it. We reduce the message overhead in route establishment by estimating the current location of the interested vehicle, and route message towards the estimated location. Simulation result shows *Ivy* is both efficient and effective, and outperforms AODV and GPSR with higher delivery ratio, lower delay and smaller message overhead.

References

[1] http://grouper.ieee.org/groups/scc32/dsrc/index.html. 5.9 GHz Dedicated Short Range Communications (DSRC).

[2] http://support.google.com/maps/bin/answer.py?hl=en&answer=61454. Traffic in Google Maps.

[3] http://msdn.microsoft.com/en-us/library/jj136866.aspx. Traffic in Bing Maps.

[4] http://www.census.gov/geo/www/tiger/. U.S. Census Bureau TIGER/Line 2009.

[5] http://sumo.sourceforge.net/. Centre for Applied Informatics (ZAIK) and the Institute of Transport Research German Aerospace Centre, SUMO – Simulation of Urban Mobility.

[6] IEEE 802.11p wireless access in vehicular environments (WAVE). In Proceedings IEEE 1609-Family of Standards for Wireless Access in Vehicular Environments (WAVE). U.S. Department of Transportation. January 9, 2006. Retrieved 2007-07-15.

[7] P. Bose, P. Morin, I. Stojmenovic, and J. Urrutia. Routing with guaranteed delivery in ad hoc wireless networks. ACM Wireless Networks, 7(6):609–616, November 2001.

[8] C.-H. Chou, K.-F. Ssu, and H. C. Jiau. Geographic forwarding with dead-end reduction in mobile ad hoc networks. IEEE Transactions on Vehicular Technology, 57(4):2375–2386, July 2008.

[9] T. Clausen and P. Jacquet. Optimized link state routing protocol (OLSR). Internet Engineering Task Force, 2003.

[10] K.-T. Feng, C.-H. Hsu, and T.-E. Lu. Velocity-assisted predictive mobility and location-aware routing protocols for mobile ad hoc networks. IEEE Transactions on Vehicular Technology, 57(1):448–464, 2008.

[11] R. Hall. An improved geocast for mobile ad hoc networks. IEEE Transactions on Mobile Computing, 10(2): February 2011.

[12] Sanqing Hu, Yu-Dong Yao, and A.U. Sheikh. Slotted aloha for cognitive radio users and its tagged user analysis. In Proceedings Wireless and Optical Communications Conference (WOCC), 2012.

[13] Sanqing Hu, Yu-Dong Yao, A.U. Sheikh, and M.A. Haleem. Tagged user approach for finite-user finite-buffer S-Aloha analysis in AWGN and frequency selective fading channels. In Proceedings IEEE 34th Sarnoff Symposium, 2011.

[14] H.-Y. Huang, P.-E. Luo, M. Li, D. Li, X. Li, W. Shu, and M.-Y. Wu. Performance evaluation of suvnet with real-time traffic data. IEEE Transactions on Vehicular Technology, 56(6):3381–3396, 2007.

[15] M. Jerbi, R. Meraihi, S.-M. Senouci, and Y. Ghamri-Doudane. Gytar: Improved greedy traffic aware routing protocol for vehicular ad hoc networks in city environments. In Pro-

ceedings 3rd ACM International Workshop on Vehicular Ad Hoc Networks (VANET), 2006.

[16] D. Johnson, D. Maltz, and J. Broch. DSR: The dynamic source routing protocol for multi-hop wireless ad hoc networks. In Proceedings Ad Hoc Networking, 2001.

[17] B. Karp and H. T. Kung. GPSR: Greedy perimeter stateless routing for wireless networks. In Proceedings Mobicom, 2003.

[18] F. Kuhn, R. Wattenhofer, Y. Zhang, and A. Zollinger. Geometric ad-hoc routing: Of theory and practice. In Proceedings of the 22nd Annual Symposium on Principles of Distributed Computing, Boston, MA, July 2003, pages 63–72.

[19] S-B. Lee, G. Pan, J-S. Park, M. Gerla, and S. Lu. Secure incentives for commercial ad dissemination in vehicular networks. In Proceedings Mobihoc 2007.

[20] I. Leontiadis and C. Mascolo. Geopps: Geographical opportunistic routing for vehicular networks. In Proceedings IEEE Workshop on Autonomic and Opportunistic Communication, 2007.

[21] C. Lochert, H. Hartenstein, J. Tian, H. Fubler, D. Hermann, and M. Mauve. A routing strategy for vehicular ad hoc networks in city environments. In Proceedings IEEE Intelligent Vehicles Symposium, 2003.

[22] X. Ma, M.-T. Sun, G. Zhao, and X. Liu. An efficient path pruning algorithm for geographical routing in wireless networks. IEEE Transactions on Vehicular Technology, 57(4):2474–2488, July 2008.

[23] Y. Mylonase, M. Lestas, and A. Pitsillides. Speed adaptive probabilistic flooding in cooperative emergency warning. In Proceedings of International Conference on Wireless Internet, 2008.

[24] V. Namboodiri and L. Gao. Prediction-based routing for vehicular ad hoc networks. IEEE Transactions on Vehicular Technology, 56(4):2332–2345, 2007.

[25] V. Naumov and T. R. Gross. Connectivity-aware routing (CAR) in vehicular ad hoc networks. In Proceedings IEEE INFOCOM 2007.

[26] C. Perkins, E. Royer, and S. Das. Ad hoc on-demand distance vector (AODV) routing. Internet Engineering Task Force, 2003.

[27] E. Perkins and P. Bhagwat. Highly dynamic destination-sequenced distance-vector routing (DSDV) for mobile computers. In Proceedings Sigcomm, 1994.

[28] Mudhakar Srivatsa, Ling Liu, and Arun Iyengar. Eventguard: A system architecture for securing publish-subscribe networks. ACM Transactions on Computer Systems (TOCS), 29(4):10, 2011.

[29] T. Taleb, E. Sakhaee, A. Jamalipour, K. Hashimoto, N. Kato, and Y. Nemoto. A stable routing protocol to support its services in VANET networks. IEEE Transactions on Vehicular Technology, 56(6):3337–3347, 2007.

[30] J. Tian, L. Han, K. Rothermel, and C. Cseh. Spatially aware packet routing for mobile ad hoc inter-vehicle radio networks. In Proceedings IEEE Intelligent Transportation Systems, 2003.

[31] H. Wu, R. Fujimoto, R. Guensler, and M. Hunter. MDDV: A mobility-centric data dissemination algorithm for vehicular networks. In Proceedings 1st ACM International Workshop on Vehicular Ad Hoc Networks (VANET), 2004.

[32] T. Yan, W. Zhang, and G. Wang. DOVE: Data dissemination to a fixed number of receivers in VANET. In Proceedings IEEE International Conference on Sensor, Mesh and Ad Hoc Communications and Networks (SECON), 2012.

[33] T. Yan, W. Zhang, G. Wang, and Y. Zhang. GOT: Grid-based on-road localization through inter-vehicle collaboration. In Proceedings IEEE International Conference on Mobile Ad hoc and Sensor Systems (MASS), 2011.

[34] Saleh Youse, Eitan Altman, Rachid El-Azouzi, and Mahmood Fathy. Analytical model for connectivity in vehicular ad hoc networks. IEEE Transactions on Vehicular Technology, 57(6), 2008.

[35] J. Zhao and G. Cao. VADD: Vehicle-assisted data delivery in vehicular ad hoc networks. IEEE Transactions on Vehicular Technology, 57(3), May 2008.

Biographies

Tan Yan is a PhD candidate in Computer Science Department in New Jersey Institute of Technology, New Jersey, USA. He received his M.E in Electrical Engineering from New Jersey Institute of Technology in 2008, and B.E. in Electrical Engineering from Southeast University, Nanjing, China in 2007.

Guiling Wang (Grace) joined NJIT in fall 2006 and was promoted to Associate Professor with tenure in June 2011. She received her Ph.D. in Computer Science and Engineering and a minor in Statistics from The Pennsylvania State University in May 2006. She received her B.S. in Software from Nankai University in Tianjin, China.

Performance Metrics for Self-Positioning Autonomous MANET Nodes

Janusz Kusyk,[1] Jianmin Zou[2], Stephen Gundry[2], Cem Safak Sahin[3]
and M. Ümit Uyar[2]

[1]*The United States Patent and Trademark Office, Alexandria, VA, USA;
e-mail: janusz.kusyk@uspto.gov*
[2]*The Department of Electrical Engineering, The City College of New York, NY,
USA; e-mail: {jzou00, sgundry00}@citymail.cuny.edu, uyar@ccny.cuny.edu*
[3]*BAE Systems – AIT, Burlington, MA, USA; e-mail: csafaksahin@gmail.com*

Received 20 May 2013; Accepted 15 July 2013

Abstract

We present quantitative techniques to assess the performance of mobile ad hoc network (MANET) nodes with respect to uniform distribution, the total terrain covered by the communication areas of all nodes, and distance traveled by each node before a desired network topology is reached. Our uniformity metrics exploit information from a Voronoi tessellation generated by nodes in a deployment territory. Since movement is one of the most power consuming tasks that mobile nodes execute, the average distance traveled by each node (ADT) before the network reaches its final distribution provides an important performance assessment tool for power-aware MANETs. Another performance metric, network area coverage (NAC) achieved by all nodes, can demonstrate how efficient the MANET nodes are in maximizing the area of operation. Using these metrics, we evaluate our node-spreading bio-inspired game (BioGame), that combines our force-based genetic algorithm (FGA) and game theory to guide autonomous mobile nodes in making movement decisions. Our simulation experiments demonstrate that these performance evaluation metrics are good indicators for assessing the efficiency of node distribution methods.

Journal of Cyber Security and Mobility, Vol. 2, 151–173.

Keywords: Topology control, MANETs, node-spreading, uniformity measures, Voronoi tessellation, area coverage, game theory, bio-inspired algorithms.

1 Introduction

Mobile ad hoc networks (MANETs) are useful for many commercial and military applications where network coverage is needed over a terrain without an established infrastructure. Autonomous topology control algorithms aim to provide a method to deploy mobile assets without a centralized controller such that MANETs are scalable and robust to node failures. In this context, it is advantageous when the MANET topology has reduced sensing overshadows, limited blind spots, enhanced spectrum utilization, and simplified routing procedures while lowering power consumed by each node. Achieving these objectives necessitate that autonomous nodes in MANETs (a) place themselves over an unknown geographical terrain in order to obtain a uniform network distribution, (b) reduce the total distance traveled before overall network objectives are reached, and (c) preserve network connectivity while attaining positions that ensure a high coverage of the area by all nodes. In this article, we present quantitative methods to evaluate performance of node self-positioning techniques with respect to uniformity of distribution of nodes, average distance traveled (ADT) by each autonomous vehicle, and the total area coverage (NAC) obtained by all nodes.

The uniform distribution of mobile nodes is often a desired network topology that helps to prolong network lifespan by ensuring that nodes deplete their energy resources evenly. When MANET nodes are uniformly distributed, they are able to equally share sensing and communication tasks, hence the likelihood that a single node ceases to function much earlier than expected due to power exhaustion is reduced. In order to gauge the performance of a MANET with respect to its uniform distribution, we define metrics based on Voronoi regions generated by the nodes.

Since a node's movement is typically the most energy-consuming operation performed by autonomous vehicles, reducing the distance that nodes travel is an important objective. ADT provides a realistic metric for evaluating self-spreading algorithms where preserving scarce energy resources is imperative.

Gathering information about operational environments to provide mission-critical data is often the main motivation for deployment of MANETs. In order to adequately utilize the communication coverage of deployed nodes, network topologies that increase NAC while preserving connectivity are often needed. Consequently, NAC provides an intuitive metric to assess how well existing communication coverage resources are utilized.

Our metrics offer design-aiding techniques for power-aware MANETs where balancing desired network performance with power-limiting constraints are imperative to maximize the utilization of deployable resources. We demonstrate the practicality of Voronoi-based uniformity, ADT, and NAC metrics by applying them to our node-spreading bio-inspired game (Bio-Game). BioGame combines our force-based genetic algorithm (FGA) and game theory (GT) concepts to guide autonomous mobile nodes in selecting locations that improve uniformity, and network coverage while limiting the distance that nodes travel. In this paper, we present a formal definition of our BioGame and compare its simulation results to the outcomes attained by mobile nodes guided by our FGA alone.

The rest of this article is organized as follows. A brief overview of the related research is presented in Section 2. Section 3 formally introduces our uniformity metrics as well as ADT and NAC evaluation techniques. In Sections 4 and 5, we define our BioGame and analyze its performance by conducting simulation experiments, respectively.

2 Related Work

In this section, we present background to Voronoi tessellation and topology control methods as well as our earlier research. An interested reader can find an extensive analysis of GT in the work by Fudenberg and Tirole [8]. Holland [13] and Mitchell [21] present the essentials of genetic algorithms (GAs).

2.1 Background

Voronoi tessellation has been applied to analyze biological cell models and the territorial behavior exhibited by animals [3, 10]. Lu et al. [19], use centroidal Voronoi tessellation for the efficient placement of vertices when rendering surfaces in computer graphics. Voronoi diagrams can also be applied to facial recognition algorithms, as demonstrated by Abbas et al. [1]. Bash and Desnoyers [2] present a distributed method for nodes in a sensor network that use Voronoi region boundaries to assist in achieving improved load balancing and energy conservation. Other Voronoi-based applications include quality measures for point distribution in an area, as presented by

Nguyen et al. [22], and optimal distribution of resources through applications of a centroidal Voronoi tessellation method, examined by Du et al. in [7].

Topology control of mobile nodes in MANETs has been studied in various contexts. In [29] and [4], nodes with a fixed configuration in a MANET dynamically adjust their power levels to achieve k-connectivity. Garro et al. [9] present a bio-inspired algorithm that allows mobile nodes to find unobstructed paths to predefined targets. Differential evolution (DE) has been successfully applied to decentralized robotic applications. In [28], Vahdat et al. present DE and particle swarm optimization that are applied to the global localization of mobile robots. In [26], DE is used for MANETs to improve the performance of routing protocols. DE is used as the mechanism for MANET nodes to choose cluster heads, as shown in [5]. Managing the movement of nodes in network models where each node is capable of changing its own spatial location has been approached by employing concepts of potential fields [14], a Lloyd-based algorithm [6], and various GA-based decentralized topology control mechanisms [24].

2.2 Our Earlier Research

In our earlier work [11, 17], we presented three distinct methods for autonomous MANET nodes to position themselves over unknown deployment areas using various GT, evolutionary GT, and FGA concepts. We introduced a force-based genetic algorithm topology control approach for uniform deployment of autonomous vehicles over a two-dimensional unknown area in [25, 27]. In [12], we introduced a differential evolution-based topology control mechanism, called TCM-DE, which we modeled as an inhomogeneous Markov chain to demonstrate its convergence towards an adequately separated final distribution of mobile nodes. We studied models that combine various GT and genetic algorithms concepts for autonomous MANET nodes positioning themselves over an unknown deployment areas in [15, 16, 30].

2.3 Contribution of This Paper

The initial concepts used to evaluate the performance metrics of self-positioning autonomous MANET nodes were introduced in [11, 18]. In this paper, we introduce the formal definition of our BioGame. Then, using our performance metrics, we formally analyze the performance of our BioGame and FGA with respect to the average distance traveled, uniform distribution of nodes, and area coverage. We verify formal analysis results with simulation experiments and show that both BioGame and FGA can be effective in providing uniform node distribution. However, BioGame, which utilizes game theory, can make better informed decisions and at the same time reduce traveling distance for nodes. This paper presents a formal definition of our BioGame together with its convergence properties. Since the nodes running BioGame base their decisions not only on the expected improvement of their own locations but also on the possible movements of their near neighbors, we are able to demonstrate that BioGame provides better informed movement decisions for the mobile nodes. Using simulation experiments, we verify that the performance of MANET nodes guided by BioGame is better than by our FGA alone.

3 Performance Evaluation Methods for MANETs

In this section, we present quantitative methods for assessing performance of MANETs with respect to the uniform distribution of mobile nodes, NAC, and ADT.

3.1 Uniformity Metrics

Equally distanced and connected mobile nodes are necessary to achieve many network goals. A uniform distribution of mobile nodes helps to simplify high-level network communication and routing operations as well as provide adequate area coverage for environment-sensing purposes. Furthermore, since the lifespan of a MANET under limited-power conditions often depends on the continuous operation of all nodes, it is important to ensure that the nodes deplete their energy resources evenly and to limit the number of nodes that cease to function prematurely. In uniformly distributed networks, where each node has the same sensing area and distance to its neighbors, power utilized by every mobile node to perform its tasks is similar and, consequently, prolonged uninterrupted operation of a MANET can be accomplished.

To gauge the performance of MANETs with respect to the uniformity of autonomous node distribution, we introduce metrics based on various quantities associated with the Voronoi tessellation [23] of the deployment terrain. Our Voronoi tessellation metric associates each node n_i with a Voronoi region V_i such that all locations that are closer to n_i than to any of the other nodes

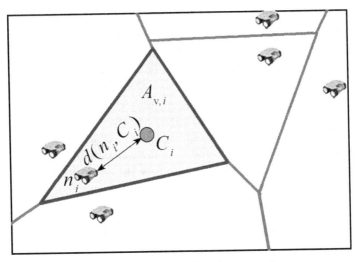

Figure 1 The Voronoi tessellation of a rectangular area.

are parts of its V_i. The Voronoi region for each MANET node is defined as

$$V_i = \{\omega \in \Omega : d(n_i, \omega) < d(n_j, \omega), \forall_{n_j \in I \setminus \{n_i\}}\} \tag{1}$$

where Ω represents the set of all positions in the deployment area, I is a set of all players (nodes), and $d(n_i, \omega)$ represents the Euclidean distance between node n_i and a location in the deployment area (i.e., (x_i, y_i) and (x_ω, y_ω)). The Voronoi tessellation of a deployment terrain is a collection of all nodes' Voronoi regions. Let the area of V_i be defined as $A_{v,i}$ and C_i be the center of mass of region V_i. Figure 1 presents a tessellation of the rectangular constant terrain depicting basic quantities associated with each Voronoi region.

In Figure 1, the darker region represents the area, $A_{v,i}$ of node n_i's Voronoi cell V_i. The parameter $d(n_i, C_i)$ denotes the distance to the center of mass of generated by node n_i's Voronoi region. For clarity of presentation, Figure 1 does not depict any values associated with the other five nodes.

We introduce two methods for measuring the uniform distribution of MANET nodes over a deployment terrain. The first metric, called $\mathcal{U}_A$, exploits differences in sizes of the areas for Voronoi regions generated by the nodes. We define $\mathcal{U}_A$ as

$$\mathcal{U}_A = \frac{1}{\bar{A}_v} \sqrt{\frac{1}{|I|} \sum_{n_i \in I} (A_{v,i} - \bar{A}_v)^2} \tag{2}$$

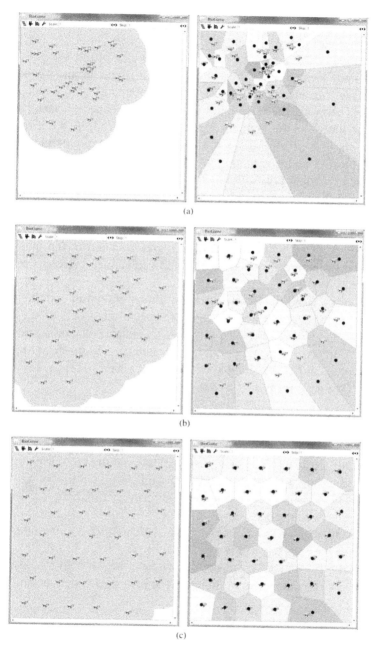

Figure 2 Examples of node distributions and corresponding Voronoi tessellations obtained by BioGame at step: (a) $t = 5$; (b) $t = 15$; and (c) $t = 50$. The center of mass for each Voronoi region is marked by a black dot.

where $\bar{A}_v$ is the arithmetic mean of $A_{v,i}$ for all $n_i \in I$ and $|I|$ denotes a total number of nodes in the network.

If mobile nodes are equally separated, the size of the Voronoi cell area for each node located in the interior of the deployment terrain is equal. Slight variations in Voronoi regions exist near the boundaries of the deployment territory. Therefore, the tessellation of the deployment area that closely resembles collection of congruent regular hexagons reflects a desirable node distribution. The metric $\mathcal{U}_A$ approaches zero as autonomous mobile agents improve their locations towards a uniform network distribution, where nodes cannot improve their positions any further.

Figure 2 shows three sample node distributions achieved by our BioGame and the Voronoi tessellations associated with them. The center of mass for each Voronoi region is marked by a black dot. The values of $\mathcal{U}_A$ for the topologies depicted in Figures 2(a), (b), and (c) are 1.6, 0.6, and 0.2, respectively, which are consistent with improvement achieved by the network at these steps.

Our second metric for network uniformity is based on the distance between the location of a node n_i and the center of mass of its Voronoi region C_i. In a given topology, the center of mass C_i indicates the preferred location for node n_i in order to best monitor its surrounding. The distance between the location of n_i and C_i indicates how close its present position is to the ideal position. The uniformity measure $\mathcal{U}_C$ is defined as

$$\mathcal{U}_C = \frac{1}{|I|} \sum_{n_i \in I} d(n_i, C_i) \tag{3}$$

where $d(n_i, C_i)$ is the Euclidean distance between the present position of n_i and the center of its Voronoi region C_i (Figure 1).

When a network converges to a uniform distribution, the separation among neighboring nodes equalizes and for all $n_i \in I$, and the distance $d(n_i, C_i)$ approaches zero. For example, the uniformity measures $\mathcal{U}_C$ for the topologies depicted in Figures 2(a), (b), and (c), are 7.8, 3.1, and 0.9, respectively. Consequently, in both of our metrics $\mathcal{U}_A$ and $\mathcal{U}_C$, the smaller value achieved by the network indicates the better placement of nodes.

3.2 Average Distance Traveled

Another important metric for assessing the performance of node self-spreading algorithms is ADT. Let s_i^t represent a strategy of player n_i at time t that corresponds to the spatial coordinates of n_i (i.e., $s_i^t = (x_i^t, y_i^t)$). Furthermore, for all $t \geq 0$, let $d(n_i^0, n_i^t)$ denote the total distance traveled by n_i up to time t. We define ADT(t) as the average distance traveled by a node until time t as

$$\text{ADT}(t) = \frac{1}{|I|} \sum_{n_i \in I} d(n_i^0, n_i^t) \tag{4}$$

In our simulation experiments, as t increases, the value of ADT never decreases (i.e., ADT is a monotonically increasing function). The rate that ADT grows is an indicator of the dynamic nature of the network. As the network reaches a uniform distribution, where nodes are not able to improve their positions any further, the derivative of ADT is zero.

3.3 Network Area Coverage

NAC is defined as a ratio of the coverage achieved by the communication areas of all nodes to the total deployment terrain. If any part of the region is covered by more than one mobile node, the overlapped area is included in NAC calculations only once. Also, only the part of node's coverage area that falls within the area of deployment counts towards NAC. Let $A_{C,i}$ denote the area covered by node n_i and A_C be the size of the area of deployment. We formally define NAC as

$$\text{NAC} = \frac{\bigcup_{n_i \in I} A_{C,i}}{A_C} \tag{5}$$

where $\bigcup$ represents the union of all coverage areas of subscribed nodes. A NAC value of one implies that the entire area is fully covered. Hence, obtaining the highest possible NAC by mobile agents is one of the goals for our game-theoretic and bio-inspired node spreading techniques.

4 Our BioGame

In our BioGame, each mobile node makes movement decisions based solely on local data. First, a node runs our novel FGA to find a set of preferred next locations and evaluates them through the spatial game set up among itself and its current neighbors. In BioGame, a set I of m players represents all active nodes in the network and for all $n_i \in I$, a set of strategies S_i stands for the possible locations into which player n_i can move. Let N_i denote the set of neighbors of node n_i in its communication range R_C, which defines n_i's sensing and communication areas. Strategy profile s for player n_i represents strategies of all nodes in $\{n_i \cup N_i\}$.

Our FGA exploits inherent characteristics of GAs, which can efficiently explore multiple possible solutions in each evolutionary step providing a set of desired solutions at the end of its computation. The fitness function used by our FGA is based on the virtual forces envisioned to be inflicted on a mobile node by its neighbors. The virtual force F_{ij} exerted on node n_i by node $n_j \in N_i$ is calculated according to the following equation

$$F_{ij} = \begin{cases} \gamma_i \left(R_c - d_{ij} \right) & \text{if } 0 < d_{ij} < d_{th} \\ \epsilon & \text{if } d_{th} \leq d_{ij} \leq R_C \end{cases} \tag{6}$$

where d_{ij} is the distance between mobile nodes n_i and n_j, d_{th} is the threshold value to define the best separation among nodes, and $\epsilon < R_c - d_{th}$. The force scaling factor γ_i is a function of the desired node degree μ and is defined as

$$\gamma_i = \frac{(|N_i| - \mu)^2 + 1}{|N_i|} \tag{7}$$

The fitness of node n_i located in s_i is influenced by its neighboring node positions represented by s_{-i}, where s_{-i} is an element in the set of all possible choices of n_i's opponents $\bar{S}_{-i}$. The fitness of n_i is calculated as

$$f_i(s_i, s_{-i}) = \begin{cases} \sum_{n_j \in N_i} F_{ij} & \text{if } N_i \neq \emptyset \\ \mathcal{F}_{\max} & \text{otherwise} \end{cases} \tag{8}$$

where $\mathcal{F}_{\max} > (m \times R_C)$ is a large penalty for mobile nodes becoming disconnected.

The fitness function in our FGA promotes a sparsely connected network topology with nodes having a limited number of neighbors and reduced overlapping communication areas when the desired number of neighbors μ is small. On the other hand, when the desired node degree is large, nodes running our BioGame will create a densely packed network, where each node has multiple neighbors, hence redundant routing paths can be established.

Figure 3 shows the surface plot of the fitness function defined by Eq. (8) for various degrees of node $n_i \in I$ and averaged distances to its neighbors in range $(0, R_c]$.

We can observe in Figure 3 that the fitness for node n_i improves when its number of neighbors approaches μ and the distance to them gets closer to R_C (i.e., nodes became spread farther apart). At any stage of the game, a player may not have the entire landscape of possible solutions to choose from, as neighbor positions may restrict it, but even a local improvement shifts the node closer to a position with the minimal virtual force inflicted on it.

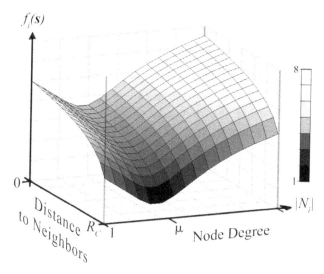

Figure 3 Fitness landscape of our FGA. For clarity of presentation, this figure does not depict the case where $N_i = \emptyset$, which results in $f_i(s) = \mathcal{F}_{\max}$.

The set of possible new locations $\bar{S}_i$ for node n_i consists of locations computed by our FGA as well as n_i's current position. Node n_i computes expected payoff for each $s_i \in \bar{S}_i$ as

$$u_i(s_i, \sigma_{-i}) = \sum_{s_{-i} \in \bar{S}_{-i}} \left(\prod_{n_j \in N_i} \sigma_j(s_j) \right) f_i(s_i, s_{-i}) \qquad (9)$$

where σ_{-i} is a probability distribution over s_{-i} and a probability of node n_j choosing location s_j is denoted by $\sigma_j(s_j)$, which represents a probability of s_j being played.

Player n_i finds the best location to move s_i^* by evaluating all elements of $\bar{S}_i$ using Eq. (9) and selecting

$$s_i^* \in \underset{s_i \in \bar{S}_i}{\mathrm{argmin}} \; u_i(s_i, \sigma_{-i}) \qquad (10)$$

that minimizes possible forces inflicted on it.

This step replaces the stochastic roulette wheel or deterministic elitism selection mechanisms in making a final decision for the new position of node n_i. However, contrary to the roulette wheel and elitism, our BioGame utilizes additional information about neighbors in order to enhance FGA performance.

In our BioGame, each node n_i autonomously determines its new location that is within R_C distance from its current coordinates.

4.1 Formal Analysis of BioGame

Let us now demonstrate that BioGame can be used by autonomous mobile nodes to efficiently disperse over an area of deployment while achieving uniform distribution and maintaining network connectivity. Let $f_i(s^t)$ represent the fitness of node n_i resulted from the strategy s_i^t being played by it at time (t). The following theorem shows that a mobile node moves to a new location only if it does not lower its fitness.

Theorem 1. *Node n_i moves to a new location s_i^{t+1} if it is expected to be better or at least as good as its present position s_i^t (i.e., $f_i(s^{t+1}) \leq f_i(s^t)$).*

Proof. Node n_i computes expected payoffs (fitness) for all of its possible next locations $\bar{S}_i$ according to Eq. (9). From the expected payoffs, node n_i selects the best location to move, as presented in Eq. (10). If there is no element in $\bar{S}_i$ for which Eq. (9) attains a smaller (i.e., better) value than for s_i^t, then $s_i^t = s_i^*$ or $u_i(s_i^t, \sigma_{-i}) = u_i(s_i^*, \sigma_{-i})$. Hence, either node n_i remains in its current position or moves to a location that gives it equally good or better expected payoff.

If, on the other hand, s_i^t is not amongst the locations that provide the minimum expected payoff for node n_i, the new location s_i^* selected by Eq. (10) must result in a better or equal expected payoff for n_i than s_i^t (i.e., $u_i(s_i^t, \sigma_{-i}) > u_i(s_i^*, \sigma_{-i})$).

Therefore, node n_i moves to a new location if and only if it has at least as good fitness as its current position.   □

We formalize the advantages of BioGame over FGA in the following theorem.

Theorem 2. *In BioGame, the decision to determine the next position for player n_i provides similar or better results than a position that is based on the outcomes of FGA only.*

Proof (sketch). Let us first assume that player n_i is the only node intending to change its location for a given moment and, consequently, no information about eventual actions of the players in N_i provide additional information for n_i. Since $\bar{S}_{-i}$ is a singleton and $\forall_{n_j \in N_i} \sigma_j(s_j^t) = 1$, where s_j^t represents the present location of player n_j, Eq. (9) becomes equivalent to Eq. (8) and, hence, player n_i selects the best new place for her to move, as ensured by Eq. (10), as if it were by using the results of our FGA only.

If, on the other hand, there is at least one other player $n_j \in N_i$ at this time intending to move according to her strategy σ_j, it is possible that this information can improve u_i selection process by using BioGame. Let $\hat{s}_i$ be the best strategy that FGA can find (either, as an outcome of elitism, roulette wheel, or similar processes), then the expected payoff resulting from moving into $\hat{s}_i$ evaluated by Eq. (9) can be at most as good as the result of Eq. (10) evaluated by our BioGame. Therefore, $u_i(\hat{s}_i, \sigma_{-i}) \geq u_i(s_i^*, \sigma_{-i})$ must hold.

As a result, player n_i can find the next best location to move by evaluating her future positions with respect to possible movements of all $n_j \in N_i$ through evaluating our BioGame. □

The following theorem illustrates that the mobile nodes running BioGame make better informed movement decisions than the nodes that make their choices regarding next positions based on FGA only.

Theorem 3. *For any given two neighboring nodes u_i and u_j, both at non-ideal locations, BioGame provides better informed movement decisions than* FGA.

Proof (sketch). Using Eq. (8), each node running FGA computes its next position regardless of possible actions of its neighbors. Therefore, it is possible that two nodes u_i and u_j may move to new locations which improve their own fitness but decrease the fitness and uniformity of the entire network. This can happen because each node only selfishly considers the improvement of its own location in its fitness calculation. Consider two nodes u_i and u_j, that are attempting to move and guided by FGA . It is possible that they may choose the same location as their target s^{t+1} (i.e., $s^{t+1} = s_i^{t+1} = s_j^{t+1}$) if the location s^{t+1} provides improvement for both nodes over their respective current positions s_i^t and s_j^t. The nodes u_i and u_j will then have to fix their positions at time $(t+2)$, as FGA will guide them to better successive positions.

However, a node running BioGame takes into account the intended decisions of neighboring nodes using Eq. (9). As long as the expected payoff is better for one node to move to a given location than the expected payoff of a node and its neighbors moving into the same location, nodes will not move into the same location. Therefore, as long as $\sigma_j(s^{t+1})f_i(s_i^{t+1}, s_j^{t+1}) > \sigma_j(s^{t+1})f_i(\hat{s}_i^{t+1}, s_j^{t+1})$, node u_i will refrain from moving to s_i^{t+1} in favor of moving to $\hat{s}_i^{t+1}$, as assured by Eq. (10). □

As a direct result of Theorem 3, we can state the following corollary regarding the performance of our BioGame.

Corollary 1. *BioGame achieves convergence faster than* FGA *for autonomous mobile nodes spreading themselves to a desired network configuration.*

Theorems 1, 2, 3, and Corollary 1 state that player u_i can improve its performance by executing BioGame to determine the best next position to move and improve the network convergence time. This observation has been further validated by the results of our simulation experiments presented in Sect. 5 below.

5 Simulation Experiments

We developed a simulation platform for our BioGame using MASON [20]. Our software implementation provides a graphical user interface allowing for a real-time visualization of ongoing network dynamics and collecting data needed for further analysis. All of our experiments were performed for MANETs with autonomous nodes determining their next locations by means of BioGame and FGA alone. To reduce noise in the collected data, each experiment was performed 20 times and the results were averaged.

For each experiment, we initially placed 40 nodes in the upper-left corner of the 100×100 units deployment area (Figure 4(a)). For simplicity and without loss of generality, all mobile agents have the same communication radius of $R_C = 16$. Our initial distribution imitates a realistic situation where the nodes enter a terrain from a common point (e.g., initiating nodes into a post-earthquake zone or a territory occupied by hostile forces) compared to random or other initial node deployments we often see in the literature. Deployed autonomous mobile nodes have no *a priori* knowledge of the underlining area and locations of their neighbors. A typical final distribution of 40 nodes running BioGame for 100 steps is shown in Figure 4(b).

5.1 NAC Improvement for Networks Running BioGame and FGA

Figure 5 shows the improvement of NAC for networks where nodes are running BioGame and FGA. In Figure 5, the vertical axis represents the ratio of the total deployment terrain covered by nodes and the horizontal axis represents the duration of the experiments. We can see in Figure 5 that mobile nodes directed by our BioGame converge faster than when mobile nodes that are directed by FGA. Also, it can be observed in Figure 5 that in the early

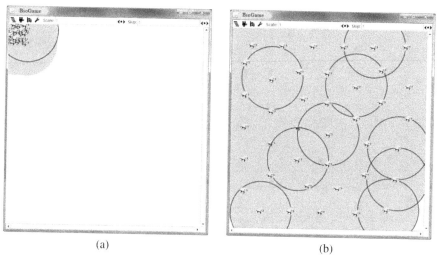

(a)   (b)

Figure 4 A typical (a) initial and (b) final node distribution of 40 nodes and their communication areas (darker color) at the beginning and end of the BioGame experiments. To better visualize BioGame performance, only the communication areas of a few selected nodes are outlined.

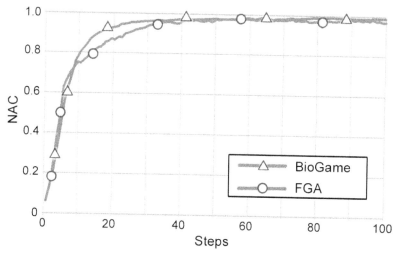

Figure 5 NACs obtained by networks running BioGame and FGA.

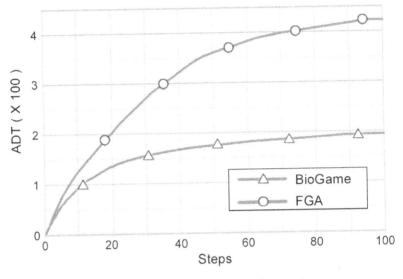

Figure 6 ADTs for a node running BioGame and FGA.

stages of the experiments, the NAC for BioGame and FGA have the highest improvement rate, indicating that the nodes are able to disperse far from their initial locations especially at the beginning of the experiments and showing effectiveness of both BioGame and FGA in finding new positions.

5.2 Average Distance Traveled by each Node

When illustrating changes in ADT for our experiments, the vertical axis represents the average total distance traveled by a node up to the time indicated in the horizontal axis. Figure 6 compares ADT for nodes running BioGame and FGA in a network consisting of 40 mobile agents. As we could observe in Figure 5, the area covered by mobile agents running FGA and BioGame are very similar. However, Figure 6 shows that the average distance traveled by a node running FGA is almost twice of that for BioGame. Specifically, Figure 6 shows that at step 35, ADT by a node running FGA is approximately 300 whereas it is about 160 for a node running BioGame. At step 50, when FGA and BioGame networks approach their maximum area coverages for this example (Figure 5), a node selecting its next location based on FGA traveled more than twice the distance of a node using BioGame (Figure 6). Conversely, by the time BioGame achieves 98% of coverage by traveling distance of approximately 160, FGA has only achieved 78% of area coverage (i.e., Figure 6, shows that FGA network ADT is 160 at step 15). The ability of BioGame to significantly reduce the required distance that nodes have to travel to accomplish predefined coverage objectives assures its practical value for all realistic implementations for which power is a scarce resource.

Another observation that we can make from Figure 6 is that ADT continues to increase throughout the experiment when mobile nodes use FGA to guide their movements. This observation shows that the nodes running FGA need more time to attain a uniform network topology than the BioGame nodes. One reason for the lower performance of FGA is that multiple nodes simultaneously may attempt to move to the same location, and delay uniform node distribution. These types of inefficient movements are greatly reduced by BioGame, since each node considers its own decisions and future actions of its near neighbors.

5.3 BioGame and FGA Uniformity Evaluation

We demonstrate the improvement in network uniformity when mobile nodes utilize BioGame and FGA to evolve towards their final distributions by using the metrics $\mathcal{U}_A$ and $\mathcal{U}_C$, which were presented in Sect. 3.1. Figure 7 shows the improvement of $\mathcal{U}_A$ and $\mathcal{U}_C$ as simulation experiments progress. We can see in Figure 7 that both BioGame and FGA converge rapidly towards a uniform distribution over the area of deployment. The largest improvement occurs during the initial iterations of the simulations showing the effectiveness of our approaches in finding new locations when the space is not constrained. However, as ADT results demonstrated, BioGame provides a more efficient method for spreading autonomous mobile agents over an unknown terrain since the nodes utilizing BioGame move less while providing better results with respect to NAC and appropriate separation among the mobile nodes (Figure 7).

6 Concluding Remarks

We presented quantitative techniques for gauging the performance of MANET nodes with respect to the uniform distribution of mobile assets, total terrain covered by communication areas of all nodes (NAC), and the distance that each node travels before a desired network topology is reached (ADT). A uniform distribution of mobile agents prolongs network's lifespan by ensuring even depletion of energy resources available to each node. We demonstrated

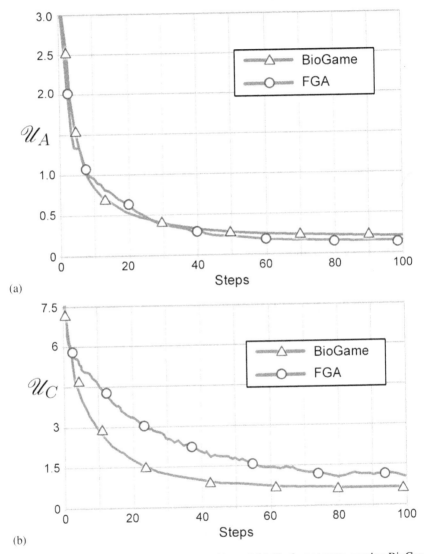

(a)

(b)

Figure 7 The improvement of uniformities (a) $\mathcal{U}_A$ and (b) $\mathcal{U}_C$ for MANETs running BioGame and FGA.

uniformity metrics that exploit Voronoi tessellations of a deployment territory to evaluate regularity in the placement of nodes. ADT can be used to measure power-efficiency of a node distribution, as movement of the nodes is one of the most power-consuming tasks. In order to adequately utilize existing mobile agents, an autonomous node self-positioning method should strive to maximize the total area covered by all nodes while preserving network connectivity. We define NAC metric as a ratio of area covered by all nodes with respect to the deployment territory. Each performance metric gives a valuable insight into the mobile network performance and collective examination of their respective results provides a comprehensive assessment of MANETs.

We present a node-spreading bio-inspired game (BioGame) combining our force-based genetic algorithm (FGA) and game theory to guide autonomous mobile agents in modeling movement decisions. Using our MANET evaluation metrics, we compare the performance of BioGame and FGA. Experimental results show that both BioGame and FGA can provide promising levels of area coverage with near uniform node distributions. However, BioGame can achieve a network topology where nodes uniformly cover the deployment terrain while traveling less than half of the distance than mobile nodes running FGA to achieve similar uniformity and NAC results. Furthermore, simulation results demonstrate that the presented metrics are useful for evaluating the performance autonomous mobile node deployment mechanisms.

Acknowledgements

Earlier versions of this work was supported by U.S. Army Communications-Electronics RD&E Center contracts W15P7T-09-CS021 and W15P7T-06-C-P217, and by the National Science Foundation grants ECS-0421159 and CNS-0619577. The contents of this document represent the views of the authors and are not necessarily the official views of, or are endorsed by, the U.S. Government, Department of Defense, Department of the Army or the U.S. Army Communications-Electronics RD&E Center.

References

[1] C. Abbas, M. Dzulkifli, and A. Azizah. Exploiting Voronoi diagram properties in face segmentation and feature extraction. Pattern Recognition, 41(12):3842–3859, 2008.

[2] B. A. Bash and P. J. Desnoyers. Exact distributed Voronoi cell computation in sensor networks. In Proceedings of the 6th International Conference on Information Processing in Sensor Networks (IPSN'07), pages 236–243. ACM, New York, 2007.

[3] J. Bernauer, R. P. Bahadur, F. Rodier, J. Janin, and A. Poupon. DiMoVo: a Voronoi tessellation-based method for discriminating crystallographic and biological protein-protein interactions. Bioinformatics, 24(5):652–658, March 2008.

[4] J. B. D. Cabrera, R. Ramanathan, C. Gutierrez, and R. K. Mehra. Stable topology control for mobile ad-hoc networks. Communications Letters, IEEE, 11(7):574 –576, july 2007.

[5] U. K. Chakraborty, S. K. Das, and U. T. E. Abbott. Clustering in mobile ad hoc networks with differential evolution. In Proceedings 2011 IEEE Congress on Evolutionary Computation (CEC), pages 2223–2228, June 2011.

[6] J. Cortés, S. Martinez, T. Karatas, and F. Bullo. Coverage control for mobile sensing networks. IEEE Transactions on Robotics and Automation, 20(2):243–255, April 2004.

[7] Q. Du, V. Faber, and M. Gunzburger. Centroidal voronoi tessellations: Applications and algorithms. Society for Industrial and Applied Mathematics Review, 41(4):637–676, 1999.

[8] D. Fudenberg and J. Tirole. Game Theory. The MIT Press, August 1991.

[9] B. A. Garro, H. Sossa, and R. A. Vazquez. Path planning optimization using bio-inspirited algorithms. In MICAI'06: Proceedings of the Fifth Mexican International Conference on Artificial Intelligence, pages 319–330. IEEE Computer Society, Washington, DC, USA, 2006.

[10] W. George and Barlow. Hexagonal territories. Animal Behaviour, 22, Part 4:876–IN1, 1974.

[11] S. Gundry, J. Kusyk, J. Zou, C. S. Sahin, and M. U. Uyar. Performance evaluation of differential evolution based topology control method for autonomous MANET nodes. In Proceedings IEEE Symposium on Computers and Communications (ISCC), pages 228–233, May 2012.

[12] S. Gundry, J. Zou, J. Kusyk, J. Zou, C. S. Sahin, and M. U. Uyar. Markov chain model for differential evolution based topology control in MANETs. In Proceedings IEEE Sarnoff Symposium, pages 1–5, May 2012.

[13] J. H. Holland. Adaptation in Natural and Artificial Systems: An Introductory Analysis with Applications to Biology, Control and Artificial Intelligence. MIT Press, Cambridge, MA, April 1992.

[14] A. Howard, M. J. Mataric, and G. S. Sukhatme. Mobile sensor network deployment using potential fields: A distributed, scalable solution to the area coverage problem. In Proceedings Distributed Autonomous Robot Systems 5, pages 299–308, 2002.

[15] J. Kusyk, C. S. Sahin, M. U. Uyar, E. Urrea, and S. Gundry. Self organization of nodes in mobile ad hoc networks using evolutionary games and genetic algorithms. Journal of Advanced Research. Elsevier, 2:253–264, July 2011.

[16] J. Kusyk, C. S. Sahin, J. Zou, S. Gundry, E. Urrea, and M. U. Uyar. Game theoretic and bio-inspired optimization approach for autonomous movement of MANET nodes. In Handbook of Optimization, from Classical to Modern Approach, Intelligent Systems Reference Library, Vol. 38, Springer, 2011.

[17] J. Kusyk, E. Urrea, C. S. Sahin, and M. U. Uyar. Game theory and genetic algorithm based approach for self positioning of autonomous nodes. International Journal of Ad Hoc & Sensor Wireless Networks. Old City Publishing, 16(1-3):93–118, 2012.

[18] J. Kusyk, J. Zou, S. Gundry, C. S. Sahin, and M. U. Umit. Techniques for performance evaluation of self-positioning autonomous MANET nodes. In Proceedings IEEE Sarnoff Symposium, pages 1–5, May 2012.

[19] L. Lu, B. Lévy, and W. Wang. Centroidal Voronoi tessellations for line segments and graphs. In Computer Graphics Forum (Eurographics Conf. Proc.), 2012.

[20] S. Luke, C. Cioffi-Revilla, L. Panait, K. Sullivan, and G. Balan. MASON: A multiagent simulation environment. Simulation, 81(7):517–527, 2005.

[21] M. Mitchell. An Introduction to Genetic Algorithms. MIT Press, Cambridge, MA, 1998.

[22] H. Nguyen, J. Burkardt, M. Gunzburger, L. Ju, and Y. Saka. Constrained CVT meshes and a comparison of triangular mesh generators. Computetional Geometry, 42(1):1–19, January 2009.

[23] A. Okabe, B. Boots, and Ks Sugihara. Spatial Tesselations: Concepts and Applications of Voronoi Diagrams, Wiley Series in Probability and Mathematical Statistics. Wiley, 1992.

[24] C. S. Sahin. Genetic Algorithms for Topology Control Problems. LAP LAMBERT Academic Publishing, February 2011.

[25] C. S. Sahin, E. Urrea, M. U. Uyar, M. Conner, I. Hokelek, G. Bertoli, and C. Pizzo. Genetic algorithms for self-spreading nodes in MANETs. In Proceedings of the 10th Annual Conference on Genetic and Evolutionary Computation (GECCO), pages 1141–1142, 2008.

[26] J. Toutouh and E. Alba. Optimizing OLSR in VANETS with differential evolution: a comprehensive study. In Proceedings of the First ACM International Symposium on Design and Analysis of Intelligent Vehicular Networks and Applications (DIVANet'11), pages 1–8. ACM, New York, 2011.

[27] E. Urrea, C. S. Sahin, I. Hokelek, M. U. Uyar, M. Conner, G. Bertoli, and C. Pizzo. Bio-inspired topology control for knowledge sharing mobile agents. Ad Hoc Networks, 7(4):677–689, 2009.

[28] A. R. Vahdat, A. N. NourAshrafoddin, and S. S. Ghidary. Mobile robot global localization using differential evolution and particle swarm optimization. In Proceedings IEEE Congress on Evolutionary Computation (CEC2007), pages 1527–1534, Sept. 2007.

[29] Y. Wang, L. Cao, and T. A. Dahlberg. Efficient fault tolerant topology control for three-dimensional wireless networks. In Proceedings of 17th International Conference on Computer Communications and Networks (ICCCN '08), pages 1–6, Aug. 2008.

[30] J. Zou, S. Gundry, J. Kusyk, M. U. Uyar, and C. S. Sahin. 3D genetic algorithms for underwater sensor networks. International Journal of Ad Hoc and Ubiquitous Computing, 13(1):10–22, 2013.

Biographies

Janusz Kusyk, Ph.D., received B.S. and M.A. degrees in Computer Science from Brooklyn College, Brooklyn, New York in 2002 and 2006, respectively, and he received Ph.D. degree in Computer Science in the Graduate Center, The City University of New York in 2012. Currently, he is a Patent Examiner at USPTO, Alexandria, VA. His research interests are in the areas of network modeling and analysis and applications of game theory and genetically inspired algorithms to wireless networks and distributed robotics.

Jianmin Zou received his B.S. degrees in both Computer Science and Chemical Engineering from Huazhong University of Science and Technology, P. R. of China in 2009. He is currently a Ph.D. candidate at the City College of New York (CCNY) of the City University of New York (CUNY). His interests include wireless mobile ad hoc networks, underwater sensor networks, biologically inspired algorithms and game theory.

Stephen Gundry received two Bachelor of Science degrees in both Engineering Science and Physics from the City University of New York at the College of Staten Island (CSI), in 2003, and a Master of Engineering degree in Electrical Engineering from the City University of New York at the City College of New York (CCNY), in 2009 and is currently a Ph.D. candidate at this institution. His interests include biologically inspired algorithms, artificial intelligence, game theory and mobile ad hoc networks.

Cem Safak Sahin, Ph.D., received his B.S. degree from Gazi University, Turkey in 1996, M.S. degree from Middle East Technical University, Turkey in 2000, and M. Phil. and Ph.D. degrees from the City University of New York in 2010, all in Electrical Engineering. Until 2004 he was an engineer at Roketsan Inc., a leading defense company of Turkey's rocket and missile research and production programs. From 2004 to 2008, he was Principal Engineer, Systems Design at Mikes Inc., a defense company specializing in Electronic Warfare Systems, working as part of a multi-national defense project in the United States. He was Senior Software Engineer from 2008 to 2010 in Elanti System. Currently, he is Senior Research Engineer at BAE Systems-AIT in Burlington, MA. His interests include wireless ad-hoc networks, bio-inspired algorithms, communication theory, multi-sensor fusion, algorithm development, artificial intelligence, machine learning, and electronic warfare systems.

M. Ümit Uyar, Ph.D., is a Professor with the Electrical Engineering Department of the City College and the Computer Science Department of the Graduate Center of the City University of New York. His interests include bio-inspired computation with applications to the mobile ad hoc networks, distributed robotics tasks and cancer chemotherapy treatment decision support systems. Dr. Uyar was the lead principle investigator for several large grants from U.S. Army and NSF to conduct research on knowledge sharing mobile agents using bio-inspired algorithms for topology control in MANETs and for a smart robot brain on FPGA which has reliable communication capabilities. Prior to joining academia, he was a Distinguished Member of Technical Staff at AT&T Bell Labs until 1993. He is an IEEE Fellow and holds six U.S. patents. Dr. M. Umit Uyar has a B.S. degree from Istanbul Technical University, and M.S. and Ph.D. degrees from Cornell University, Ithaca, NY, all in electrical engineering.

Detecting Targeted Attacks by Multilayer Deception

Wei Wang, Jeffrey Bickford, Ilona Murynets, Ramesh Subbaraman, Andrea G. Forte and Gokul Singaraju

AT&T Security Research Center, New York, USA;
e-mail: {wei.wang.2, jbickford, ilona, ramesh.subbaraman, forte, gokul.singaraju}@att.com

Received 15 May 2013; Accepted 15 July 2013

Abstract

Over the past few years, enterprises are facing a growing number of highly customized and targeted attacks that use sophisticated techniques and seek after important company assets, such as customer data and intellectual property. Unlike conventional attacks, targeted attacks are operated by experts who use multiple steps to gain access to sensitive assets, and most of time, leave very few network traces behind for detection. In this paper, we propose a multi-layer deception system that provides an in depth defense against such sophisticated targeted attacks. Specifically, based on previous knowledge and patterns of such attacks, we model the attacker as trying to compromising an enterprise network via multiple stages of penetration and propose defenses at each of these layers using deception based detection. Due to multiple layers of deception, the probability of detecting such an attack will be greatly enhanced. We present a proof of concept implementation of one of the key deception methods proposed. Due to various financial constraints of an enterprise, we also model the design of the deception system as an optimization problem in order to minimize the total expected loss due to system deployment and asset compromise. We find that there is an optimal solution to deploy deception entities, and even over spending budget on more entities will only increase the total expected loss to the enterprise. Such a system

Journal of Cyber Security and Mobility, Vol. 2, 175–199.

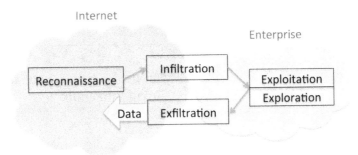

Figure 1   A multi-stage attack with layers of penetration.

can be coupled with existing detection techniques to protect enterprises from sophisticated attacks.

Keywords: Deception, honeypot, honeynet, optimization.

1 Introduction

Recent trends indicate that enterprises today face a growing threat of sophist-icated attackers who seek to steal or compromise proprietary information and assets [19, 11, 2, 1]. These attacks are executed in multiple stages and each stage is highly customized for each targeted enterprise. Attackers typically leave few network and system level footprints in attempts to evade detection. They exploit zero-day vulnerabilities to deliver malware using a single event, such as a carefully crafted spear-phishing email, as the entry point into an enterprise network. Detecting such events by leveraging existing techniques is difficult, especially when social engineering techniques are used to infilt-rate the target organization. Once an attacker has infiltrated an organization, detecting other phases of an attack, such as data exfiltration, is a very difficult process which requires correlation between multiple events, such as firewall, IDS and DNS logs. As attackers customize their attacks for individual targets, prior knowledge such as blacklists of malicious domain names, drop servers IPs and malware signatures may not be useful for the attack. Therefore, rapid detection is technically difficult and stopping an attack in the early stage is challenging.

Figure 1 illustrates a multi-stage attack with layers of penetration as an attack example of such. A typical pattern observed in these attacks is that an attacker first studies the targeted enterprise during a "reconnaissance" phase, gathering information such as the organization background, resources and individual employees to initially target to launch the attack. By using social engineering techniques, such as a spear-phishing email, the attacker attempts to "infiltrate" into the enterprise by using a particular employee as the entry point. This typically requires an employee to fall victim to the social engineering attack, for example by following a web link or opening an attachment that contains some exploit and malicious payload. During this phase of "exploitation", the attacker penetrates a level deeper by gaining control of the employee's personal assets (such as email and personal computer). This may then be used to penetrate another level deeper into the enterprise through manual "exploration" of remote servers (hosting databases, proprietary algorithms, intellectual properties etc.), or to launch additional social engineering attacks against other employees who have access to the information that the attacker seeks to obtain. Some attacks may exploit and gain control of many different servers and machines during the exploration phase to gain a persistent foothold in the enterprise. Once an asset has been obtained, the attacker finally "exfiltrates" the data out of the enterprise network and the attack can be considered successful.

This pattern, as mentioned above, reveals that there are three layers of penetration – a human layer, a local asset layer, and a global asset layer. Each layer of penetration brings the attacker closer to the targeted information assets. The human layer is the information of an enterprise employee, which is researched and gained by an attacker in the reconnaissance activity. The local asset layer refers to the employee's local machine containing immediate assets and the global asset layer represents the assets hosted on servers accessed and shared by multiple users. To address this problem, we propose to apply a multilayer deception system in order to protect important assets within an enterprise network. That is, our proposed multilayer deception system provides *detection via deception* at these three layers. By detection via deception, we are referring to the idea of placing bogus facilities and resources within a network or file system such that when accessed, an alarm is triggered and the attackers presence is discovered. In order to trick an attacker to access the item, these bogus resources are generated to appear as valuable as normal. By providing deception at each stage of an attack, the proposed system greatly enhances the chance of detecting intrusions in an early stage.

In this paper, we introduce concepts of honey people, honey files, honey servers, and honey activity to defend the human, local, and global assets layers respectively. Honey activity can be considered as fake file system or network activity intended to prevent a sophisticated attacker from evading bogus resources by observing actual user behaviors. We integrate all of these into a complete system that work cooperatively together to provide an in depth solution against targeted attacks. We also present an optimization based method to design a budget conscious defense solution which minimizes the expected loss due to asset compromise.

The following features of our work represent significant contributions towards the goal of detecting multi-stage attacks:

- Multiple layers of deception are designed specifically for each layer of penetration, so that each deception layer can increase the chance of detecting attacks at an early stage.
- By estimating the cost of deploying deception entities and asset values, proposed optimization model can find an optimal solution to intelligently allocate the budget on necessary deployment to minimize the overall expected loss. To our knowledge, this is the first proposal for an optimal system design based on the cost of deception.

The rest of the paper is organized as follows. Related work will be presented in Section 2. The details of our system are at described in Section 3. In Section 4, we describe a prototype implementation of honey files with honey activity. Section 5 presents our deception system design method using optimization and shows results for a specific example. Section 6 shows the future work to enhance the system implementation in the cloud and further study on system design optimization. Section 7 concludes our paper.

2 Related Work

This basic concept of deception has a long history of successfully being used in security. Generally a deception system consists of resources that appear to be a part of a network, but are actually isolated and monitored. The system will seem, to the intruder, to contain information or a resource of value to attackers, but such information is of no value in content. Because it is not a real entity, any actions on this resource are suspicious by nature [21, 3]. The concept of deception was firstly introduced by Clifford in his book [22], which described in detail how he hunted over months for a computer hacker who broke into a computer at the Lawrence Berkeley National Laboratory. By trapping the hacker in the decoy system, the hacker's actions were recorded and his behavior was successfully revealed and studied. By dedicating a chapter on "Traps, Lures, and Honey Pots" in their book [10], Cheswick and Bellovin discussed how to use unused services as decoys on firewalls and during another chapter "An Evening with Berferd" they logged and interacted with a hacker in order to understand an attempted attack into their network. The use of fake files as bait to detect malicious file accesses was first proposed in [25,26]. Such files can be set up by users and only those who are intimately familiar with file system can potentially avoid such a trap. Along the same line, Fiedler [13] proposed honeypots for database protection, where he described basic honeypot architecture to secure a SQL database server. This SQL database provides a honeypot trap for the intruder while still allowing the web application to run as normal. Alternatively, Bruce [20] proposed to embed macros in fake word or pdf documents to trigger alerts when files are opened. Younghee et al. [17] proposed a software-based decoy system that generates believable Java source code which, to an adversary, appears to be entirely valuable and proprietary. A honeynet is used to monitor a large and/or more diverse networks in which one honeypot may not be sufficient [7]. A honeynet can be utilized as a part of the network intrusion detection system. Instead of utilizing a variety of physical systems, Honeynet Project introduced virtual Honeynets to run honeypots on a single computer [18]. By using virtualization techniques, the Honeynet project runs several honeypots of multiple operating systems types on a single computer for analysis purposes. Project Honey Pot [15] employed a web based honeypot network which uses software embedded in web sites to collect information about IPs harvesting e-mail addresses for spam, bulk mailing and fraud. Most recently on the mobile communications front, Collin [16] implemented HoneyDroid, a smartphone honeypot for the Android operating system to catch attacks originating from the Internet, mobile networks, as well as through malicious applications; while Wang [24] introduced fake contacts on mobile devices to quickly detect messaging-based malware propagation in cellular networks. Brian et al. [8] proposed trap-based defense mechanisms and a deployment platform for addressing the problem of insiders attempting to exfiltrate and use sensitive information. Malek et al. [5] modelled user search patterns as well as touch interactions with decoy documents to detect deviations, indicating an attack.

Unlike all previous work that considers one flavor of honeypot as a system, we propose multiple layers of deception that work cooperatively. We focus on sophisticated and highly customized attacks against enterprises targeting their most valuable information assets via three levels of penetration. For each layer of penetration, we propose related methods of deception to detect such penetration. By considering multiple deception layers as one integrated system, we model the enterprise's expected losses when assets are compromised, and formulate the problem of minimizing the overall expected loss using our deception system while meeting the budget constraints as an optimization problem. There is no previous work that considered such a systematic design.

3 Multilayer Deception System

3.1 Multilayer Deception System

In this section, we describe the key concepts of our proposed multilayer deception system consisting of Honey People (*HP*), Honey File with Honey Activity (*HFHA*), and Honey Servers with Honey Activity (*HSHA*). The framework can be naturally extended to more layers of deception if required, such as Honey Smartphone Contacts and Honey Databases. Figure 2 illustrates our proposed system with these mutliple layers of deception. The system is comprised of both real entities typically used by employees and deception entities used to detect an intrusion within the network. The alerts generated from accessing the deception entities will be sent to an analyst server, where an analysis process will handle the alerts. Throughout the rest of this paper we consider only three levels of deception, though in a full-fledged deployment we would rely on previous implementations of additional deception layers.

3.2 Honey People

The goal of Honey People is to protect employees against social engineering attacks coming from outside the enterprise, such as spear phishing messages containing a URL or malicious attachment. A HP is similar to the profile of a real employee but contains bogus identity and/or contact information (e.g. multiple email addresses). These HP can possibly be posted on public web-sites (corporate web-page and popular social network sites[1]) and on physical entities such as business cards, as shown in Figure 3. By this means, HP confuses an attacker with fake information so that when the attacker chooses a target to send a phishing message to, there is a probability that a HP is chosen. A message sent to a HP is forwarded to the analyst server to ascertain whether it was a penetration attempt. If not, the message can be forwarded to the actual recipient. Detecting phishing messages is out of scope of this paper but can be relied on existing techniques [6, 9].

---

[1] Subject to terms and conditions of the sites.

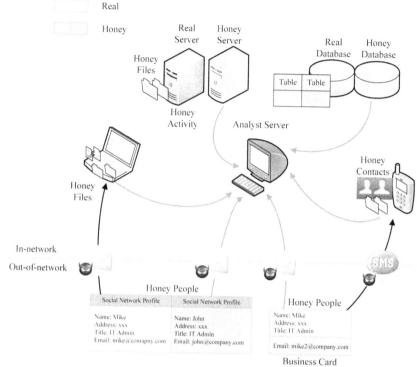

Figure 2 A multilayer deception system (H represents *honey entity* and R represents *real entity*).

The overhead at the analyst server is clearly dependent on the amount of actual emails the person receives. We believe that internally within a company, HP is not a scalable solution as employees will likely send each other a lot of emails, which is why in our system, HP is only used to protect against social engineering attempts from the outside. However, even then, for a given employee, HP may or may not be a feasible solution, or only be a partial solution depending on the employee's level of public exposure and the amount of external emails he receives. For example, a company's CEO's identity may be well known, rendering a bogus identity useless, but his email address may not, in which case HP via multiple email addresses may be feasible. On the other hand, a manager may have both a well known identity and contact address, so new clients can reach him, in which case HP is not feasible as a solution at all. In our overall system design in Section 5,

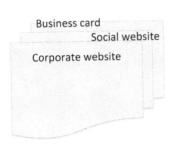

Name: Mike
Address: XYZ
Title: IT Administrator
Email: mike1@xyz.com

Name: Alex
Address: XYZ
Title: IT Administrator
Email: alex1@xyz.com

Business card
Social website
Corporate website

Figure 3   Honey people hosted on different publicly accessible medium. Mike is a real employee and Alex is a deception entity.

we capture this aspect of HP by having different costs for a HP solution for different employees.

3.3 Honey Files with Honey Activity

If the attacker manages to avoid HP defenses, an employee can potentially be successfully social engineered and infected with malware. After the attacker infiltrates the enterprise network and compromises a machine, he can begin the next stage of an attack, where local emails, files, and folders are explored, user name and password are sniffed, and even user daily activities are monitored. In the attempt that the attacker starts to compromise the employee's local assets, we introduce another layer of deception which utilizes honey files (HFs) with honey activity (HA) as a defense. Honey files are bogus files and folders that, to the attacker, are indistinguishable from real files and folders. Since the attacker, in our model, is extremely sophisticated, we assume he can obtain complete access to a machine and hence has the ability to monitor employee file and folder access behaviors. Thus, he can easily ignore files and folders that the employee never accesses, defeating the purpose of generic honey files. To obscure the view of the system, our honey files are augmented with local honey activity which updates file meta data (such as file size, name, date) to emulate actual employee accesses on real files.

In order to be convincingly realistic to the attacker, honey files should be created in separate, as well as the same, directories as the ones employees typically work with. Honey files should also be generated for files that contain evidence of attacks, such as log files. These files are typically modified or deleted by attackers in order to remove the evidence of their presence. Alerts to the analyst server can be generated upon content related operations such as read, open, move, copy, and delete. An implementation of honey files with honey activity is described in Section 4.

3.4 Honey Server with Honey Activity

If the attacker is not caught while tinkering with an employee's local assets, he may then attack global assets that the employee has access to. At this layer, our system uses honey servers with honey activity to detect the attacker's presence. A honey server is a bogus remote server mirroring real remote servers in which the employee has access to. Again since we assume the attacker is sophisticated enough to observe an employee's remote access history, he can readily avoid traditional honey servers if they are never accessed. Hence, we augment user machines with remote honey activity to emulate user network behavior, such as connecting to a server with proprietary data.

Honey activity, with either honey files or honey servers, is a means to generate activities that appears to be generated by a regular user, and performs all tasks that a real user would perform. Honey activities need to be completely indistinguishable from a real user to avoid the possibility of malware distinguishing between HA and real user activities. There are two types of HA, local and remote honey activities. The local HA generates activities staying within the machine, such as creates, updates local honey files in order to make them correlated with real files. The remote HA generates activities going out of the machine which emulates user network behaviors such as connections to data servers. In the case where an attacker follows the local HA to access updated honey files, or remote HA to establish a SSH connection to a remote server, an alert will be triggered.

In order to differentiate between honey activity and an attacker's activity to a remote server, we need to define new algorithms. In particular, the honey server and honey activity module could agree on what network patterns the honey server expects to see as a result of the honey network activity. One such agreement could be the time pattern by which the honey activity module connects to the honey server. When an attacker tries to connect to the honey server at the wrong time, the honey server will be able to identify this network activity as originating from an attacker and thus trigger an alert to the analyst server.

3.5 Analyst Server

The Analyst Server is a center where alerts are received from different deception entities and analysis is applied to confirm or remove alerts. As we mentioned before, existing techniques can be applied to detect phishing emails or messages [9,6]. Since we capture the suspicious emails or messages with full content, a live sample of the malware can be potentially obtained from either the attachment or drive-by-download link. In such a way, we can analyze the malware itself. Different sophisticated detection schemes, such as behavioral analysis, automated URL browsing, and content based detection can be applied [4,12], to check whether a piece of malware is present on a web page. Once a phishing message is confirmed, a signature can be generated and applied in the network to block future delivery.

The analyst server is also a centralized place to correlate different alerts to increase the knowledge of a penetration attempt. For example, if alerts happen on honey file entities on two different machines, then it worth comparing host applications, logs and network traffic from these machines to identifying the cause of the similar anomalies. It also could be an alert that is triggered on a honey file entity and later on anoreother alert on a honey people entity. With some analysis, these two separate alerts could be correlated to the same attack campaign.

3.6 False Positives

In general, reducing the false positives is one of the most challenging tasks in deception systems. At one hand, deception entities should be indistinguishable from real entities in order to be convincingly realistic to the attacker. On the other hand, convincingly realistic honey entities may confuse legitimate users.

There are several ways to reduce false alerts triggered by legitimate operations on protected assets. Since people are creatures of various habits, normal routine activity from a user typically follows a detectable access pattern. In a case of an enterprise laptop, a user often goes to specific workspaces, use known applications and creates new files and updates known files for his own knowledge. Not so often, a user will go to an unknown folders and manipulate unknown files. Thus, one important aspect to help reduce false positives is employee awareness. Employees would be educated on the system and trained to not perform operations on unknown files, folders, and servers.

On the technical front, for honey files and folders, names can be differentiators between honey entities and real entities. When a file is registered with the deception system, it can generate multiple honey files associated with a real file, with similar names for honey files but different identifiers. A secondary channel (such as SMS) can be utilized to deliver the identifier of the real file name. In the case of a machine being compromised, an attacker will not know the real file name unless he compromises the secondary channel at the same time. Automatic algorithms can also reduce false positives such as the time pattern described between the honey activity and honey server. A secondary channel can also be utilized to send alerts if honey entities are accessed, legitimate users can remove alerts if they operate on honey entities.

4 Implementation

Developing the multilayer deception system is ongoing work and as an initial proof of concept, we implement a system which can protect local assets using our deception approach. More specifically, the prototype focuses on protecting important files located on a user's machine. Based on a set of important documents (assets), typically located in various directories on a machine, we generate multiple *honey files* corresponding to a single asset. These files currently have the same name as the protected asset but with some identifier at the end. For example, a file called `secret.txt` will be transformed into multiple files e.g., `secret-1.txt`, `secret-2.txt`, `secret-3.txt`, etc. with `secret-2.txt` being the real file. This is currently a manual process and therefore the user must know which identifier corresponds to the actual asset. We are looking at automating the honey file generation process and maintaining a secondary channel in order for the user to identify the real asset.

When a honey file is generated, it is registered with a system level service which has the ability to monitor the file system and trigger *alert events* when a honey file is accessed. Due to the ubiquitous nature of the Windows XP operating system in enterprise corporations today, we built our prototype as a system level Windows service using C# and the .NET framework. Figure 4 shows the implementation of our system. Our *deception service* runs with administrator privileges and cannot be disabled by normal users. It is important to note that malware which exploits a vulnerability and gains administrator rights could disable our file monitoring service. If this occurs, the malware could access all honey files without triggering an alert event. To protect against these attacks, the deception service could be implemented using a hypervisor-based approach if the highest level of security is required.

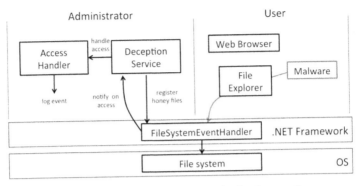

Figure 4   Kernel level deception service implementation.

The DeceptionService registers honey files with a FileSystemEventHandler, which invokes a handler function when files are accessed. When a honey file is accessed, this handler function passes the file off to an AccessHandler which maintains a queue of access events to process. While there are files within this queue, the AccessHandler thread currently logs the honey file access for later inspection. In practice, the AccessHandler can perform any task the multilayer deception system requires. In the full system implementation, we plan to notify the analyst server that the specified honey file was accessed. At this point, since we assume the user's machine is compromised, we can validate if the honey file was accessed accidentally or not through a secondary channel such as SMS. If the user determines that they did not intentionally access the honey file, the computer has been compromised and must be cleaned or replaced.

5 Deception System Design by Optimization

In practice, enterprises only have a finite budget to implement their security solutions. The available budget must be utilized in the most effective way possible. In this section we model the enterprise's expected losses when assets are compromised, and formulate the problem of minimizing the overall expected loss using our deception system while meeting the budget constraints as an optimization problem. Here, we focus on two deception layers which are honey people and honey files with honey activity, but the method can be easily extended to more layers.

5.1 Model

Consider an enterprise that has N_L local assets, N_G global assets, and M employees with different levels of direct access to these assets. Also, the employees are connected to each other via a social network and consequently, they have indirect access to all other employees' assets. The probability p_{ik}^{sn} represents the probability of user k being successfully compromised by the direct social connection from i to k if user i is compromised. We attempt to find the total probability Q_{ik}^{sn} introduced by all paths from i to k in the social network. Let node i be user i and the weight of a directed link from i to k be p_{ik}^{sn}. We then formulate this as a reliability problem in graph theory, where the objective is to calculate the reliability of node pairs in the graph, given every link may fail with a probability. A modified version of Don's algorithm [23] is applied to calculate Q^{sn} for all user pairs. Algorithm 2 recursively finds all successful paths from user i to k, which is plugged in the main algorithm 1 to calculate Q_{ik}^{sn}.

Data: Adjacency matrix for the social network A
Result: Q_{ik}^{sn} from user i to k
Initialize an empty set L, the max hop h_m;
while *h not exceed max hop h_m and A^h is not zero* **do**
    Find the paths between i and k by calculating A^h;
    Store paths in set L;
end
Initialize $Q_{ik}^{sn} = 0$;
Initialize an empty successful path set W;
if *L is not empty* **then**
    Find the shortest path l in L;
    Store path l in W;
    Execute algorithm SuccessPath(l, W, L);
    for *each path l in W* **do**
        $Q_{ik}^{sn} += \prod_{\forall link \in l}$ probability(link is success/fail);
    end
end

Algorithm 1: Calculate social impact Q_{ik}^{sn} from user i to k.

Suppose that the attacker attacks employee i with probability P_i, gets access to employee's asset j with probability p_{ij}^{ua} and gains access to employee k with probability p_{ik}^{sn}. Let L_{ij} be the loss if the attacker accesses asset j of employee i. Then the enterprise's expected total loss is defined as a sum of

Data: A path set l
Result: successful path set W
SuccessPath(l, W, L);
for *each link l_k in l* **do**
    Fail the link l_k;
    Generate a new path set L' from L by removing the failed link l_k;
    if *new path set L' is not empty* **then**
        Find the shortest path ll in the new path set L';
        Store the shortest path ll and failed link l_k in W;
    end
    SuccessPath(ll, W, L');
end

Algorithm 2: Function SuccessPath(l, W, L) to calculate the successful path set W.

direct and indirect losses, $T_{AL}(i) = D_{AL}(i) + I_{AL}(i)$, where the direct loss is represented as

$$D_{AL}(i) = P_i \left(\sum_{j=1}^{N_L} p_{ij}^{ua} L_{ij} + \sum_{j=1}^{N_G} p_{ij}^{ua} L_{ij} \right) \tag{1}$$

and the indirect loss introduced by social connections is

$$I_{AL}(i) = P_i \sum_{k=1, k \neq i}^{M} \left(\sum_{j=1}^{N_L} Q_{ik}^{sn} p_{kj}^{ua} L_{kj} + \sum_{j=1}^{N_G} Q_{ik}^{sn} p_{kj}^{ua} L_{kj} \right). \tag{2}$$

Q_{ik}^{sn} is the total probability of an attacker to compromise user k via all social paths from user i to user k.

As described in the early section, if a honey entity is attacked in an attempt to reach a real entity, an alert is raised revealing the attacker's presence. Thus, honey entities reduce the probability of an attacker compromising a real entity. We also note that social engineering could be one of the ways that an attacker can infiltrate an enterprise network. Thus, given that user i is protected by x_i^u (integer) number of honey people, the probability that the attacker compromises user i reduces to $P_i^H(x_i^u) = P_i(\alpha_1 f(x_i^u) + \alpha_2)$, where α_1 is the probability of being attacked via social engineering and α_2 by other ways ($\alpha_1 + \alpha_2 = 1$). Similarly if local asset j of a user i is protected by x_{ij}^a (integer) number of honey files with honey activity, the probability of asset j being compromised reduces to $p_{ij}^{uaH}(x_{ij}^a) = p_{ij}^{ua} g(x_{ij}^a)$, and for global assets we have $p_{ij}^{uaH}(x_j^a) = p_{ij}^{ua} g(x_j^a)$. Functions $f()$ and $g()$ should be decreasing functions of their arguments to make sure that probabilities of compromise decrease when when entities are protected by honey people and honey files. Since the attacker is equally likely to pick any of the real corresponding honey people/files, we assume that $f(z) = g(z) = 1/(z + 1)$ hereafter.

We would like to find the optimal allocation of HP(s) and HF(s) with HA that minimize the total loss in case of an attack. Therefore, the optimization problem can be formulated as:

$$\min \sum_{i=1}^{M} T_{AL}(i) =$$

$$\min_{x_{ij}^a, x_i^u, x_j^a} \sum_{i=1}^{M} \left[P_i^H(x_i^u) \left(\sum_{j=1}^{N_L} p_{ij}^{uaH}(x_{ij}^a) L_{ij} + \sum_{j=1}^{N_G} p_{ij}^{uaH}(x_j^a) L_{ij} \right) \right. \tag{3}$$

$$\left. + P_i^H(x_i^u) \sum_{k=1, k \neq i}^{M} Q_{ik}^{sn} \left(\sum_{j=1}^{N} p_{kj}^{uaH}(x_{kj}^a) L_{kj} + \sum_{j=1}^{N} p_{kj}^{uaH}(x_j^a) L_{kj} \right) \right]$$

subject to

$$\sum_{i=1}^{M} C_i^u x_i^u + \sum_{i=1}^{M} \sum_{j=1}^{N_L} C_{ij}^a x_{ij}^a + \sum_{j=1}^{N_G} C_j^a x_j^a \leq B$$

C_i^u is the cost of deploying a single HP on user i, C_{ij}^a is the cost of deploying HF on local asset j corresponding to user i's direct access to j, C_j^a is the cost of deploying a global honey asset and B is a budget on implementation of the deception system. Due to implementation restrictions, the variables, x_i^u, x_{ij}^a, and x_j^a, may need to satisfy additional constraints:

$$x_i^u \in \{0 \ldots S_i^u\}, \quad x_{ij}^a \in \{0 \ldots S_{ij}^a\}, \quad x_j^a \in \{0 \ldots S_j^a\} \quad \forall i, j$$

where S_i^u, S_{ij}^a, and S_j^a are the (integer) upper bounds on the number of HPs and HFs respectively.

5.2 Example

The following example demonstrates how the optimization model works. Assume a data analytics company have ten employees ($M = 10$), that is CEO, manager, system administrator (SysAd) and seven data analysts (DAs) as in

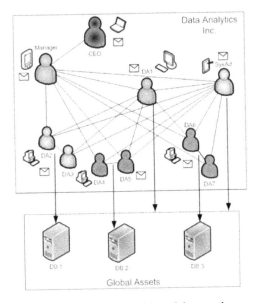

Figure 5 Local and global assets with social network connections.

Figure 5. Each employee has two local assets ($N_L = 2$): a personal computer and e-mail account. The company also has three global assets ($N_G = 3$) which are databases with different levels of access for different employees. The database 1 (DB 1) is more valuable than database 2 (DB 2), which in turn is more valuable than database 3 (DB 3). The employees have varying levels of (direct) influence on other employees in the social network. The employees and their (direct) access and (direct) influence profiles are described below.

- *CEO* has high value local assets; no access to global assets and among employees communicates only with the manager.
- *Manager* has moderate value local assets; no access to global assets; high social influence on other employees; and is highly influenced by the CEO, System Administrator and Data Analyst 1, but not others.
- *System Administrator* has low value local assets; moderate access to all global assets; high social influence on other employees, except CEO; and moderately influenced by the Manager, but not others.
- *Data analyst 1* has low value local assets; moderate access to all global assets; high social influence on the Manager and some the other Data Analysts; is highly influenced by the Manager and System

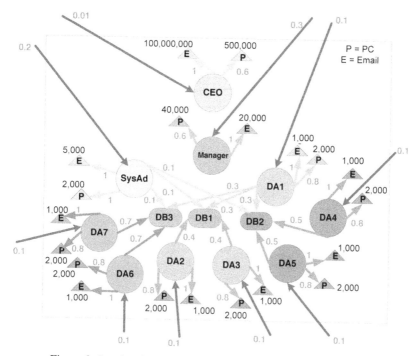

Figure 6   Local and global asset values with access probability p_{ij}^{ua}.

Administrator; and moderately influenced by some of the other Data Analysts.

- *Data analysts 2-7* have low value local assets; high access to some global assets; high influence over some employees and highly influenced by some employees.

Tables 1(a)–(f) show losses that an attacker causes by compromising each of the company employees, social influence of employees by their co-workers, attack probabilities and costs of honey people and honey files. Figures 6 and 7 represent the graph view of the parameters and values in the tables. The probability of attacks from social engineering is $\alpha_1 = 0.25$. Here, honey emails accounts and honey databases can be implemented in the system using honey files.

The optimal solution of the problem was found using Mathematica's Differential Evolution algorithm [14] for different budgets. This algorithm is a stochastic optimization method that minimizes an objective function by modelling the problem's objectives while incorporating constraints. Simil-

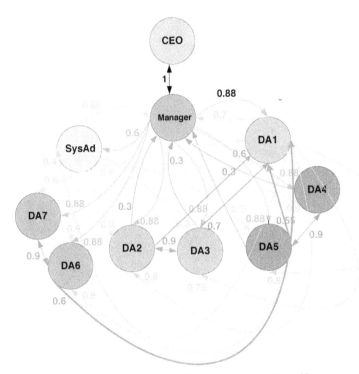

Figure 7   Social influence probability p_{ik}^{sn}.

arly to genetic algorithms, Differential Evolution algorithm is a population based algorithm using crossover, mutation and selection operators. The main steps of the Differential Evolution algorithm are initialization and evaluation followed by recursively repeating mutation, recombination, evaluation and selection steps until a termination criteria are met.

For practical reason, we limit the maximum number of honey email accounts and HP to be five, which are two conditions applied to Equation 3. There are several observations that we learn from the results.

1. If the company does not implement the deception system ($B = 0$), the expected total loss in case of attack is $122 million. If there is a certain amount of the budget available, for example, the budget of the company on implementation of honey entities is $B = \$500,000$, the deception system decreases the expected loss from $122 million to $10 million. The optimal allocation of HP is shown in Table 2(a). This solution has the following explanation: the CEO, System Administrator and Manager

Table 1 (a) Losses in million dollars from employee compromise; (b) Probabilities that an asset is compromised; (c) Matrix of (direct) social influence; (d) Attack probabilities; (e) Cost of HP (f) Cost of HF.

	E-mail	PC	DB 1	DB 2	DB 3
CEO	10	5	0	0	0
Manager	2	4	0	0	0
Sys Admin	0.5	0.2	30	15	10
DA1	0.1	0.2	15	7.5	5
DA 2	0.1	0.2	30	0	0
DA 3	0.1	0.2	30	0	0
DA 4	0.1	0.2	0	15	0
DA 5	0.1	0.2	0	15	0
DA 6	0.1	2.2	0	0	10
DA 7	0.1	0.2	0	0	10

(a)

	E-mail	PC	DB 1	DB 2	DB 3
CEO	1	0.6	0	0	0
Manager	1	0.6	0	0	0
Sys Admin	1	1	0.1	0.1	0.1
DA 1	1	0.8	0.3	0.3	0.3
DA 2	1	0.8	0.4	0	0
DA 3	1	0.8	0.4	0	0
DA 4	1	0.8	0	0.5	0
DA 5	1	0.8	0	0.5	0
DA 6	1	0.8	0	0	0.7
DA 7	1	0.8	0	0	0.7

(b)

	CEO	Manager	Sys Admin	DA 1	DA 2	DA 3	DA 4	DA 5	DA 6	DA 7
CEO	0	1	0	0	0	0	0	0	0	0
Manager	1	0	0.6	0.88	0.88	0.88	0.88	0.88	0.88	0.88
Sys Admin	0	0.85	0	0.91	0.9	0.9	0.9	0.9	0.9	0.9
DA 1	0	0.7	0.4	0	0.8	0.75	0	0.8	0.8	0
DA 2	0	0.3	0.4	0.6	0	0.9	0	0	0	0
DA 3	0	0.3	0.4	0.7	0.9	0	0	0	0	0
DA 4	0	0.3	0.4	0	0	0	0	0.9	0	0
DA 5	0	0.3	0.4	0.55	0	0	0.9	0	0	0
DA 6	0	0.3	0.4	0.6	0	0	0	0	0	0.9
DA 7	0	0.3	0.4	0	0	0	0	0	0.9	0

(c)

	CEO	Manager	Sys Admin	DA 1	DA 2	DA 3	DA 4	DA 5	DA 6	DA 7
P_i	0.25	0.35	0.2	0.1	0.1	0.1	0.1	0.1	0.1	0.1

(d)

	CEO	Manager	Sys Admin	DA 1	DA 2	DA 3	DA 4	DA 5	DA 6	DA 7
C_i^u	2 500	2 500	750	750	750	750	750	750	750	750

(e)

	E-mail	PC	DB 1	DB 2	DB 3
C_j^a	1 000	2 000	10 000	10 000	10 000

(f)

have high probabilities of being attacked. That is why they are highly protected by HP. The CEO and Manager's e-mail account and PC are highly protected, since they have very a high value for the company and could be attacked. All the databases are protected by HFs.

2. If the company has more budget to implement more entities, the expected losses will decrease as seen in Figure 8(a). But if the company aims at minimizing the total expected expenses, defined as the cost of implementation of the honey system as well as the expected losses in case of attacks, the more protection does not mean less loss. If the budget of the company on honey entities is $B = \$1,800,000$, the expected loss drops to

Table 2 Optimal solution for (a) $1,500,000 budget and (b) $1,800,000 budget.

	HP	HF_{Email}	HF_{PC}
CEO	2	5	9
Manager	4	5	8
Sys Admin	1	4	3
DA 1	5	2	1
DA 2	3	2	3
DA 3	4	2	3
DA 4	5	1	3
DA 5	5	0	4
DA 6	2	1	1
DA 7	5	1	3

HF_{DB1}	HF_{DB2}	HF_{DB3}
14	12	8

(a)

	HP	HF_{Email}	HF_{PC}
CEO	5	5	21
Manager	4	5	23
Sys Admin	3	5	12
DA 1	3	5	24
DA 2	5	5	4
DA 3	5	5	7
DA 4	5	5	2
DA 5	5	5	10
DA 6	5	3	5
DA7	5	4	9

HF_{DB1}	HF_{DB2}	HF_{DB3}
56	38	44

(b)

$4.5 million. In fact, this budget minimizes the total expected expenses. This optimal solution is shown in Table2(b). This budget protects local assets and global databases even more, with highest level of protection on email accounts for all employees.

3. For the budget exceeding the optimal number $1,800,000, the total expected expenses increase which means over deploying honey entities is not necessary, since improvements in the expected loss will not cover the additional budget spent as shown by Figure 8(b).

6 Future Work

As enterprises move their assets into cloud infrastructure, the ability to deploy and manage a multi-layer deception approach becomes much more realistic. We are currently looking at implementing a multi-layer deception approach within a cloud-based enterprise network. The use of Virtual Desktop Infrastructure (VDI) to access company assets, hosted on some managed shared storage device, allows the possibility of implementing honey files from outside of the potentially infected desktop machine. In this case, honey files can be created and monitored from the shared storage device itself or using virtual machine introspection techniques to ensure that an attacker does not subvert this layer of deception. This also leads to a single location for honey file management instead of managing every individual employee machine located within the enterprise.

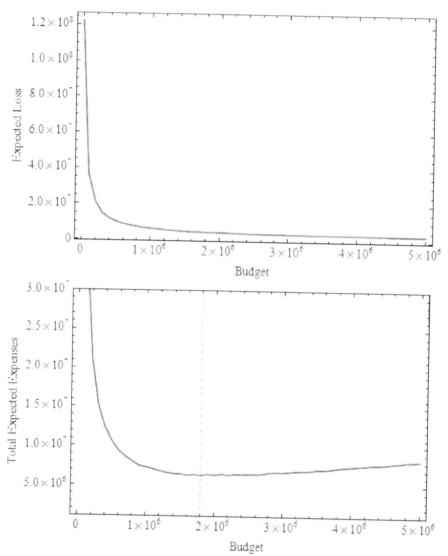

Figure 8   (a) Expected loss as a function of budget (in dollar units), (b) Expected total expenses as a function of budget (in dollar unit). The vertical line represents the optimal budget.

Within an enterprise cloud, honey servers can be easily implemented using cloned virtual machines of servers hosting ligitimate services. In this case, if an attacker scans the internal network for vulnerable servers, the honey server would automatically have the same services and fingerprint of the legitimate server. To the employees of the enterprise, these cloned servers may have slightly different domain names such that the user never travels to them.

In this type of implementation, the ease of spawning honey entities results in a tunable meter for honey entity generation. For example, during normal operation there may be a one to one mapping of honey servers throughout the important server assets within the enterprise network. If at some point in time one of the layers of deception is triggered, for a window of time after a potential compromise, an increased number of honey entities could be easily spawned throughout the other layers of deception in order to increase the probability of catching the attacker. The cost of this approach and its impact on employee productivity must be studied further during our future work.

Another piece of the future work is to better estimate the cost of deploying honey entities and values of each asset within the enterprise network. A survey will be planed to estimate these values, and we will run the optimization algorithm on a real world case to help validate our optimization result in this paper.

7 Conclusion

In this paper, we propose a multilayer deception system that provides in depth defense against sophisticated attacks. We propose defenses at each layer that an attacker may target, via deception based detection. The fact that multiple layers of deception are applied, the probability of detecting the presence of an attacker early is greatly enhanced. Furthermore, a mathematical optimization model is utilized to decide what deception entities should be deployed on which assets to minimize the total expected loss if being attacked, given a limited budget.

As future work, we plan to focus on system implementation in the cloud environment, in particular, we will study interactions between honey activities and honey servers. Also, we will implement a secondary channel to identify real file names and notify user to confirm alerts. In the optimization work, we will study more scenarios with different assumptions on the targeted assets.

References

[1] `www.wired.com/threatlevel/2011/06/citi-credit-card-breach`, 2011.

[2] Night dragon. `blog.industrialdefender.com/?p=725`, 2011.

[3] Edward G. Amoroso. Cyber Attacks: Protecting National Infrastructure. Elsevier Science, 2010.

[4] Ulrich Bayer, Imam Habibi, Davide Balzarotti, Engin Kirda, and Christopher Kruegel. A view on current malware behaviors. In Proceedings of the 2nd USENIX Conference on Large-scale Exploits and Emergent Threats: Botnets, Spyware, Worms, and More, LEET'09, pages 8–8, Berkeley, CA, USA, USENIX Association, 2009.

[5] Malek ben Salem and Salvatore J. Stolfo. Modeling user search behavior for masquerade detection. In Proceedings of the Fourteenth Symposium on Recent Advances in Intrusion Detection, 2011.

[6] Andre Bergholz, Jan De Beer, Sebastian Glahn, Marie-Francine Moens, Gerhard Paass, and Siehyun Strobel. New filtering approaches for phishing email. Journal of Computer Security, 18:7–35, 2010.

[7] Jagjit S. Bhatia, Rakesh Sehgal, Bharat Bhushan, and Harneet Kaur. Multi layer cyber attack detection through honeynet. In New Technologies, Mobility and Security, (NTMS'08), pages 1–5, 2008.

[8] Brian M. Bowen, Shlomo Hershkop, Angelos D. Keromytis, and Salvatore J. Stolfo. Baiting inside attackers using decoy documents. 2009.

[9] Madhusudhanan Chandrasekaran, Krishnan Narayanan, and Shambhu Upadhyaya. Phishing e-mail detection based on structural properties. In Proceedings NYS Cyber Security Conference, 2006.

[10] William R. Cheswick. An evening with Berferd, in which a cracker is lured, endured, and studied. In Proceedings of the USENIX, January 1992.

[11] Damballa. The command structure of the aurora botnet. `http://www.damballa.com/research/aurora/`, March 2010.

[12] Manuel Egele, Theodoor Scholte, Engin Kirda, and Christopher Kruegel. A survey on automated dynamic malware-analysis techniques and tools. ACM Computing Surveys (CSUR), 44(2):6, 2012.

[13] C. Fiedler. secure your database by building honeypot architecture using a sql database firewall. `http://archive.is/o1TW`.

[14] Yuelin Gao, Zaimin Ren, and Yang Gao. Modified differential evolution algorithm of constrained nonlinear mixed integer programming problems. Information Technology Journal, pages 2068–2075, 2011.

[15] Project HoneyPot. A web based honeypot network. `projecthoneypot.org`.

[16] Collin Mulliner, Steffen Liebergeld, and Matthias Lange. Honeydroid – Creating a smartphone honeypot. Technical report, Technische Universität Berlin, 2011.

[17] Younghee Park and Salvatore J. Stolfo. Software decoys for insider threat. In 7th ACM Symposium on Information, Computer and Communications Security, 2012.

[18] Honeynet Project. Know your enemy: Defining virtual honeynets. `old.honeynet.org/papers/virtual`, 2003.

[19] RSA. RSA security brief: Mobilizing intelligent security operations for advanced persistent threats, 2011.

[20] Bruce Schneier. www.schneier.com/blog/archives/2011/11/fake_documents.html.
[21] Lance Spitzner. Honeypots: Tracking Hackers. Addison-Wesley Longman Publishing, Boston, MA, 2002.
[22] Clifford Stoll. The Cuckoo's Egg. Doubleday, New York, 1989.
[23] Don Torrieri. An efficient algorithm for the calculation of node-pair reliability. In Proceedings IEEE Military Communication Conference, 1991.
[24] Wei Wang, Ilona Murynets, Jeffrey Bickford, Christopher Van Wart, and Gang Xu. What you see predicts what you get – Lightweight agent based malware detection. Wiley Journal, Security and Communication Networks, 2012.
[25] J. Yuill, M. Zappe, D. Denning, and F. Feer. Honeyfiles: Deceptive files for intrusion detection. In Proceedings from the Fifth Annual IEEE SMC Information Assurance Workshop, pages 116–122, 2004.
[26] Jim Yuill, Dorothy Denning, and Fred Feer. Using deception to hide things from hackers: Processes, principles, and techniques, 2006.

Biography

Wei Wang finished her Ph.D. degree from Stevens Institute of Technology in 2010. Now she is a Member of Technical Staff in AT&T Security Research Center. Her research interests are mainly in data analysis, intrusion detection and prevention, and applying machine learning techniques in network security, especially in mobile networks.

Jeffrey Bickford is a researcher with the Chief Security Office at AT&T. He is currently completing his M.S. at the Rutgers University Department of Computer Science. He is interested in mobile device security with a focus on using virtualization techniques to create a secure and robust mobile platform. Prior to joining the Security Research Lab he was a summer intern at AT&T Research in Florham Park.

Ilona Murynets is a scientist at the Chief Security Office at AT&T. She obtained her Ph.D. in Systems Engineering, Stevens Institute of Technology. Ilona holds B.Sc. degree in Mathematics and M.S. degree in Statistics, Financial & Actuarial Mathematics from Kiev National Taras Shevchenko University, Ukraine. Ilona's research is in the area of data mining, optimization and statistical analysis in application to fraud detection, malware propagation, mobile and network security.

Ramesh Subbaraman is a Member of Technical Staff at the AT&T Chief Security Office's Security Research Center. His research interests are in communication network design and architecture, networking protocols design & analysis, network data mining & analytics, and network security. In addition to traditional approaches, he is very interested in using principles from mathematical optimization, machine learning and mechanism design in networking.

Andrea G. Forte is a Researcher within the Chief Security Office at AT&T. He earned both his Master's Degree and Bachelor's Degree in Telecommunications Engineering at the University of Rome "La Sapienza" in Italy. The paper based on his dissertation work on fast handoffs for real-time media in IEEE 802.11 wireless networks was commercialized by SIPquest Inc. His research interests include mobility, real-time media, location-based services, wireless networks and Internet of Things.

Gokul Singaraju is a developer in AT&T Chief Security Office. His previous experience includes Motorola, NEC Laboratories America, Eulix Networks Inc., Hughes Network Systems, Indotronix International Corp, and Texas Instruments India Ltd. He received Masters in Technology (Computer Science) Indian Statistical Institute, Kolkata 1994.

Online Manuscript Submission

The link for submission is: www.riverpublishers.com/journal

Authors and reviewers can easily set up an account and log in to submit or review papers.

Submission formats for manuscripts: LaTeX, Word, WordPerfect, RTF, TXT.
Submission formats for figures: EPS, TIFF, GIF, JPEG, PPT and Postscript.

LaTeX

For submission in LaTeX, River Publishers has developed a River stylefile, which can be downloaded from http://riverpublishers.com/river_publishers/authors.php

Guidelines for Manuscripts

Please use the Authors' Guidelines for the preparation of manuscripts, which can be downloaded from http://riverpublishers.com/river_publishers/authors.php

In case of difficulties while submitting or other inquiries, please get in touch with us by clicking CONTACT on the journal's site or sending an e-mail to: info@riverpublishers.com